nendo

nendo

Foreword

At the *Homo Faber* exhibition, my eye was caught by a beautiful nendo vase.

It was on display in the seventeenth-century Longhena Library, one of the most fascinating rooms at the Cini Foundation in Venice, where a discerning selection of original vases was curated by Milan's Triennale Design Museum. The aim was to tell the story of design, of its changing styles and evolving tastes, through the medium of flower vases. This was a very interesting idea, something of a pilot exhibition that still had plenty of scope for development. Vases are ancient history. Five thousand years ago, in the city of Ur in Babylonia, terracotta amphorae were already being produced on an 'industrial' scale. The functional as well as aesthetic evolution of these artefacts also provides a clear and simple picture of how the taste and meaning of craftsmanship have moved through the ages.

This extensive introduction brings us to the work of nendo, which lies deep in the furrow of that evolving design and is not just an aesthetic feat. Against the background of the Venice exhibition, their Patchwork-Glass vase exuded an instant thrill. A work of highly skilled craftsmanship, it is made up of numerous interconnected snippets of glass. Whether these were welded, glued or fired together in the furnace is hard to tell. But the mystery of the object's technical manufacture makes its beauty still more intriguing. The result is a marvel. Moreover, it implies that even such a plain and humble, mature and experimented-on object as a vase can still be reinvented and turned into something different. This, I believe, is what design today is all about: it's about injecting new energy into things that have always existed, experimenting with untested approaches and creating new forms to represent the anthropological and imaginative needs of contemporaneity.

The nendo studio is a master in this respect. All of its designs are fanciful, bold and innovative. Its expressive flair is both poetic and parsimonious, making its objects aesthetically delightful yet rooted in formal, functional, structural and material innovation. Often completely white or formed by subtle lines, they are never overstated; they have no superstructures or pointless decorative features. Their essentiality immediately reveals an evolutionary urge and suggests that, luckily, there are still a whole lot of things to be invented. This makes us happy. For it's a quality not easily found in the works of other designers.

I have a deep admiration for Oki Sato. His nendo studio has perfectly absorbed the message of daring, freakish fancy and revolutionary desire delivered by Memphis. But also that of its immediate successor, the minimalist phenomenon and its search for purity, essentiality and dematerialization. I shall always be grateful to Oki for his friendly interest and curiosity during a visit to my office in Milan.

Michele De Lucchi, October 2018

2002–5

Canvas

Naja

A project to transform an old house into a restaurant, using just one sheet of canvas. With a 200-metre (656-foot) length of fabric, it was possible to wrap the entire building, both inside and outside, as well as to upholster the furnishings. Any offcuts of the material were used to create business cards, matchboxes and menus. While the use of a single material was necessary for keeping costs down, it also allowed a unified aesthetic to be maintained throughout the design.

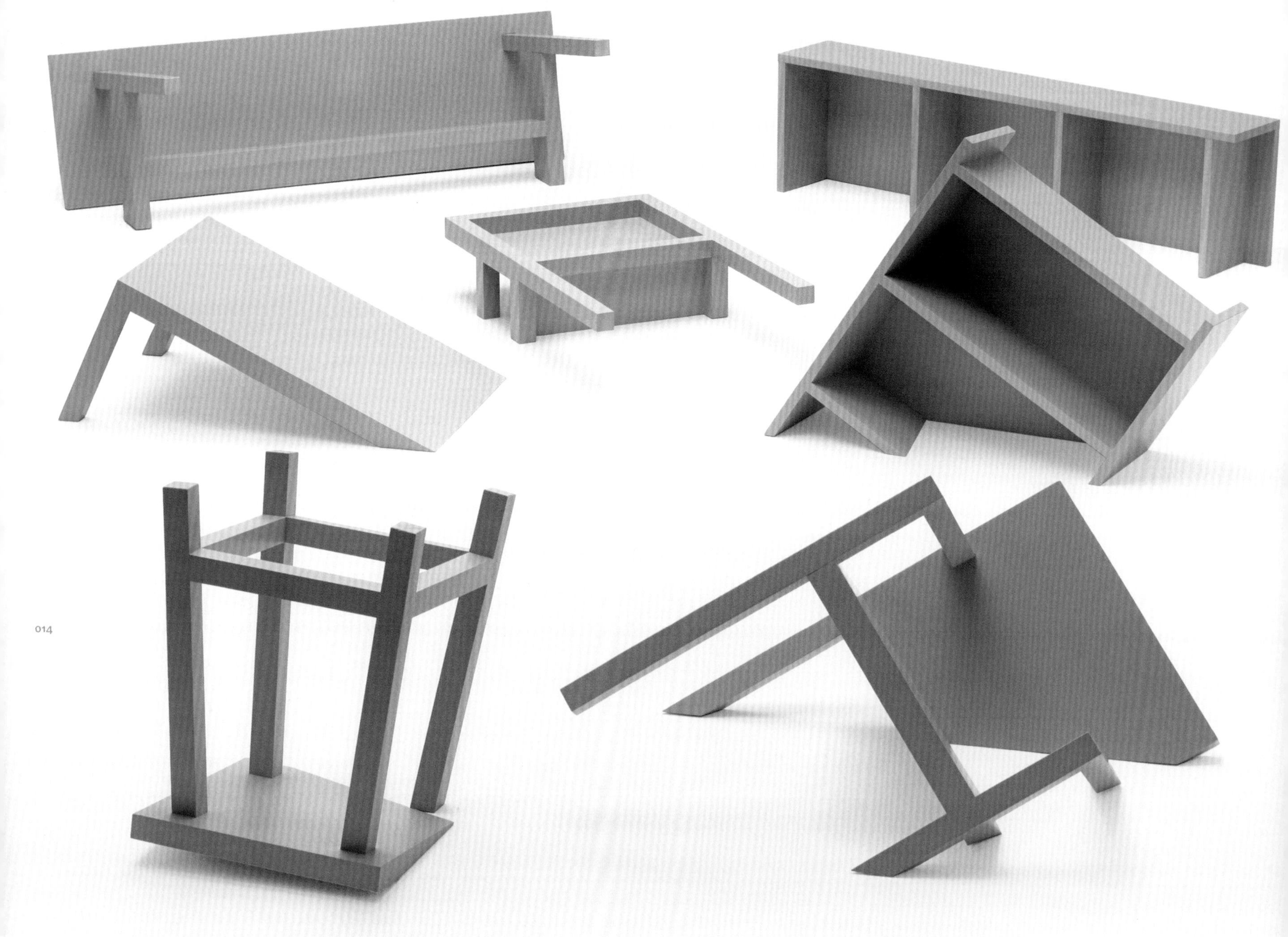

Sinking About Furniture

Tokyo Design Week

In designing a collection of furniture that appears to be sinking into the floor, the aim was to encourage a reconsideration of the relationship between form and function. A table with one leg submerged, a side table toppled over on to its side, slowly disappearing into the unknown: this series showed how design, through losing its original purpose, can bring about new and unique forms of use.

Rakuyo-Bench

Tokyo Designers Block

When sitting on this bench, *rakuyō-* (fallen leaf-) shaped lights appear on the ground below, slowly changing colour from green to red. When leaving, the lights gradually fade away, as if the leaves are decomposing. The effect is intended to reflect the passing relationship between strangers and how the traces of one's existence are rarely noticed by others.

Drawer House

Private Commission

A compact living space with domestic features – such as beds, a bathroom, stairs, shelving and a television – concealed inside its walls. These can be pulled out when needed, just like drawers in a piece of furniture. Another influence in Drawer House's design was the relationship between a computer and its software: the space is only 'initialized' once it has been 'formatted', or organized, by the user. Individuals are free to design each room as they wish; rather than sleeping in a bedroom, for example, the place where the occupant sets the bed becomes the 'bedroom'. When there is more than one occupant, the space may then be subdivided by pulling out the shelving unit. This simple yet flexible system is very effective in the context of Tokyo's restricted housing situation.

020

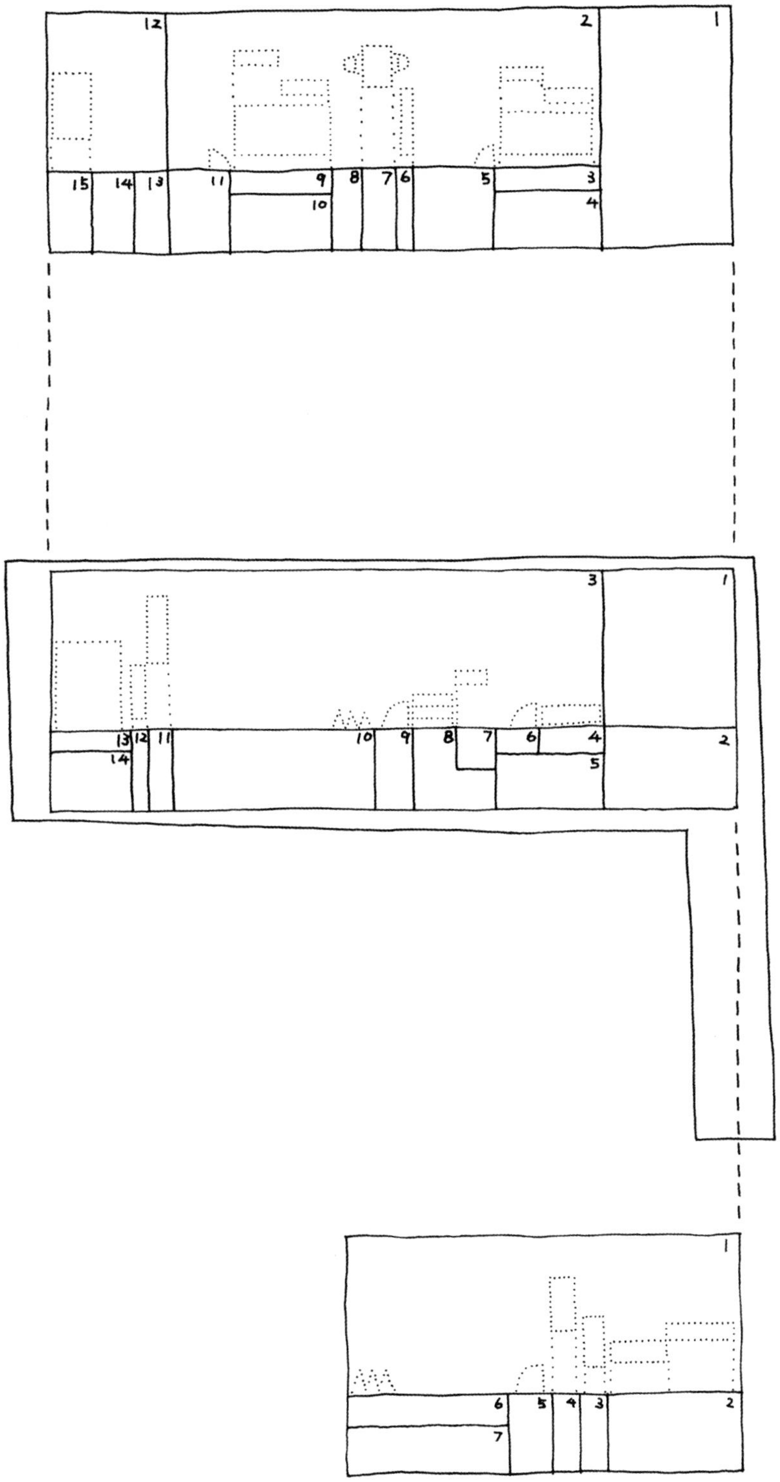

2F

1. VOID
2. LIVING SPACE
3. STORAGE
4. BED
5. STAIRCASE
6. BOOKSHELF
7. DESK/CHAIR
8. WARDROBE
9. STORAGE
10. BED
11. TOILET
12. TERRCE
13. DRESSING SPACE
14. BATHROOM
15. BATH TUB

1F

1. GARDEN
2. SKYLIGHT
3. LIVING SPACE
4. SHOE BOX
5. ENTRANCE
6. STAIRCASE
7. POWDER ROOM
8. STAIRCASE
9. TOILET
10. KITCHEN
11. TV
12. CUP BOARD
13. BED
14. PANTRY

BF

1. LIVING SPACE
2. BED
3. STORAGE
4. STORAGE
5. CABINET
6. CORRIDOR
7. STAIRCASE

Wind

Swedese

As if it were a piece of cloth billowing in the breeze, the Wind stool was designed for maximum structural performance using only the minimum number of elements. Such a balance was achieved through the use of ultra-thin plywood and carbon-fibre moulding.

Uzu

Cibone

A simple spiral forms the basis of the Uzu coat stand, allowing hats and garments to be hung up and held in place between its rings. Starting at the wider end of the spiral, users slide their coat collars between the gaps and around to the narrower section, which secures the items in place. Made from a single steel pipe that was formed through bending, and with no welding required, Uzu was inspired by the basic shape of a spring.

Key-Calendar

Idea International

A calendar that operates through a series of keyholes, each one representing a different day of the week, month and date. A specific date is indicated by inserting a series of three keys, an action that encapsulates the feeling of opening a new door every day.

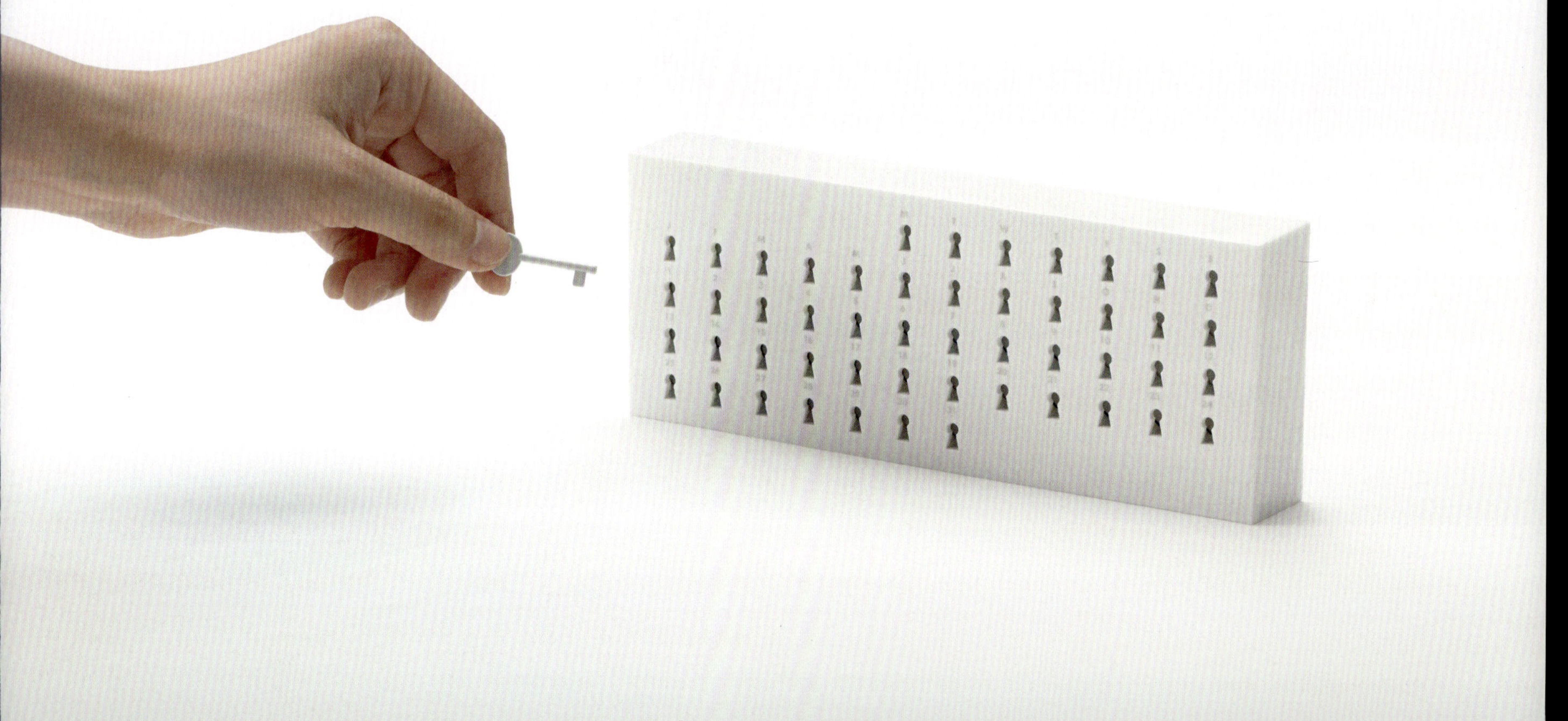

Kazan
H Concept

Unlike traditional tissue dispensers that double up as containers for the tissues themselves, Kazan's hollow, volcano-shaped form instead sits on top of a stack of tissues, functioning as a kind of paperweight to hold them in place. By pulling the topmost tissue through the hole in the weight's centre, it creates a shape similar to a plume of volcanic smoke. The aim with the design was not 'how to conceal', but rather 'how to show' its contents.

Book House

Private Commission

A combination house and library with bookshelves cladding its exterior walls, so protecting the privacy of its inhabitants' living space while inviting visitors to interact with the books that surround it. Book House is located on Shikine-jima island, to the south of Tokyo, where there was previously no library. Now, the island's inhabitants can bring and return books freely. Semi-transparent fibre-reinforced plastic (FRP) divides the interior from the exterior spaces, allowing daylight to filter through between the bookshelves during the day, and letting the interior light out at night. This interplay of light, which changes daily according to the arrangement of books on its shelves, creates a connection between the dwelling's internal and external environments.

ラモーナとおとうさん
ひとりっ子エレンと親友
飛ぶ教室
宝島
ラスムスくん英雄になる
My Day

Karaoke-Tub

Moc

A karaoke booth inspired by the bathroom. Originating from the idea that everybody loves to sing in the shower, the design features a folding shower door that leads to a low, tub-like sofa, surrounded by a white tiled floor and showerhead 'coat hangers' that line the walls – each element adding to the relaxed and playful atmosphere. Turning on the red 'tap' emits warm-coloured ambient light, while turning on the blue 'tap' boosts the flow of cool-coloured illumination.

A sense of oneness with others is easily evoked in public bathing areas like Japan's hot springs and municipal baths; in such places, the things that normally provide people with privacy are removed. Taking a bath together means deepening a relationship through trust, as does this Karaoke-Tub.

Hie-Hie
Top Runner

A range of glasses that have the appearence of already containing chilled water – an effect that was achieved through sandblasting and frosting. A silkscreen-printing technique was also used to add a transparent dewdrop-type pattern. The series includes four types of glasses, including a collins and a tumbler, and each glass displays three different levels of frosting.

Chab-Table

De Padova

The Japanese *chabudai* is a multipurpose, short-legged table that can easily be moved, folded or stored away when not in use. The Chab-Table draws inspiration from the flexibility of the traditional *chabudai* whilst also building on the design to add functionality. Its separate leg frames can be used horizontally or vertically, allowing two different table heights. In a high position, it can be used as a side table; in a low position, as a coffee table – and, in removing the legs altogether, it can even serve as a tray.

2006–7

Hanabi

Milan Design Week

The increasing temperature of the bulb makes the 'petals' of this shape-memory alloy lamp open – and therefore 'bloom' – whenever the light is turned on, gently closing again once it is turned off. The literal translation of *Hanabi* (fireworks) is 'flower of fire'. Both flowers and fire have an essentially temporary nature. Like its dual namesakes, this light flickers between a vibrant presence and inevitable disappearance.

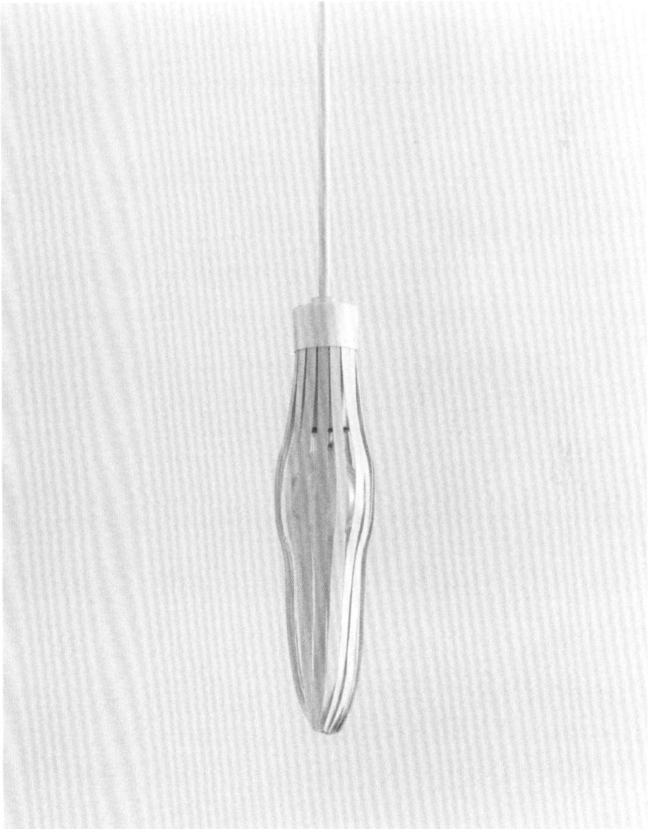

Polar
Milan Design Week

A series of three glass nesting tables, which can be stacked, partially overlapped or lined up side-by-side. Each table top contains a patterned sheet of polarizing film so that when they are juxtaposed, flowers appear through their previously clear surfaces. It is this film that gives the tables their name, and also because the resulting floral motifs appear as though trapped in ice.

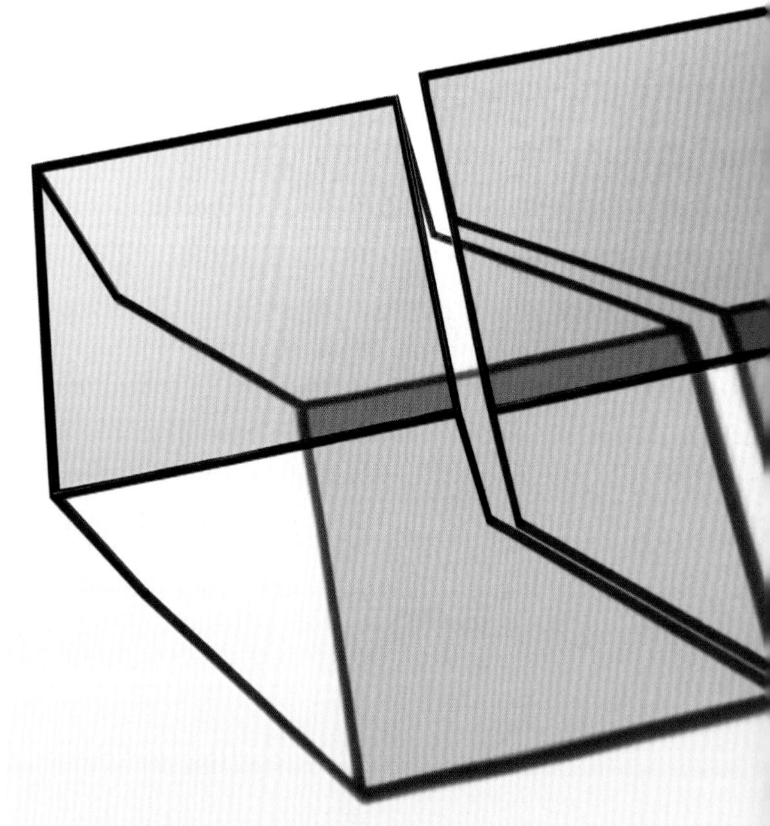

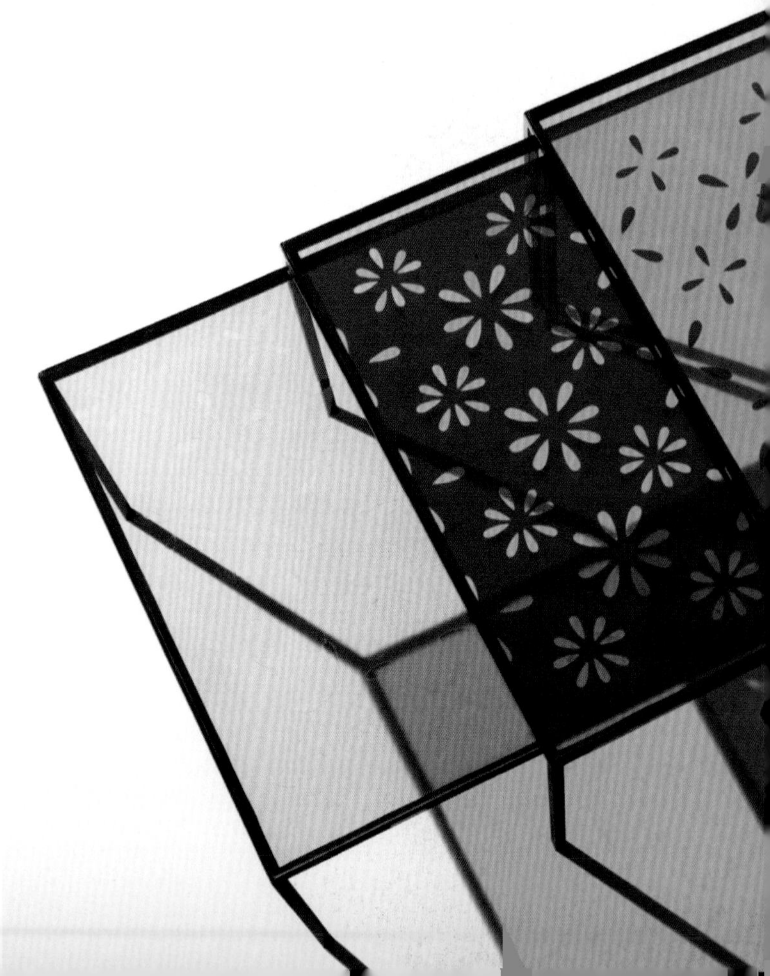

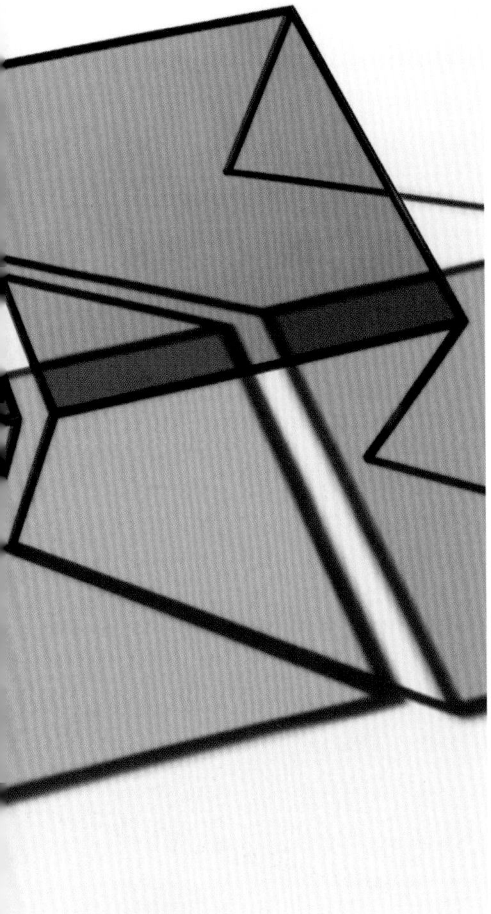
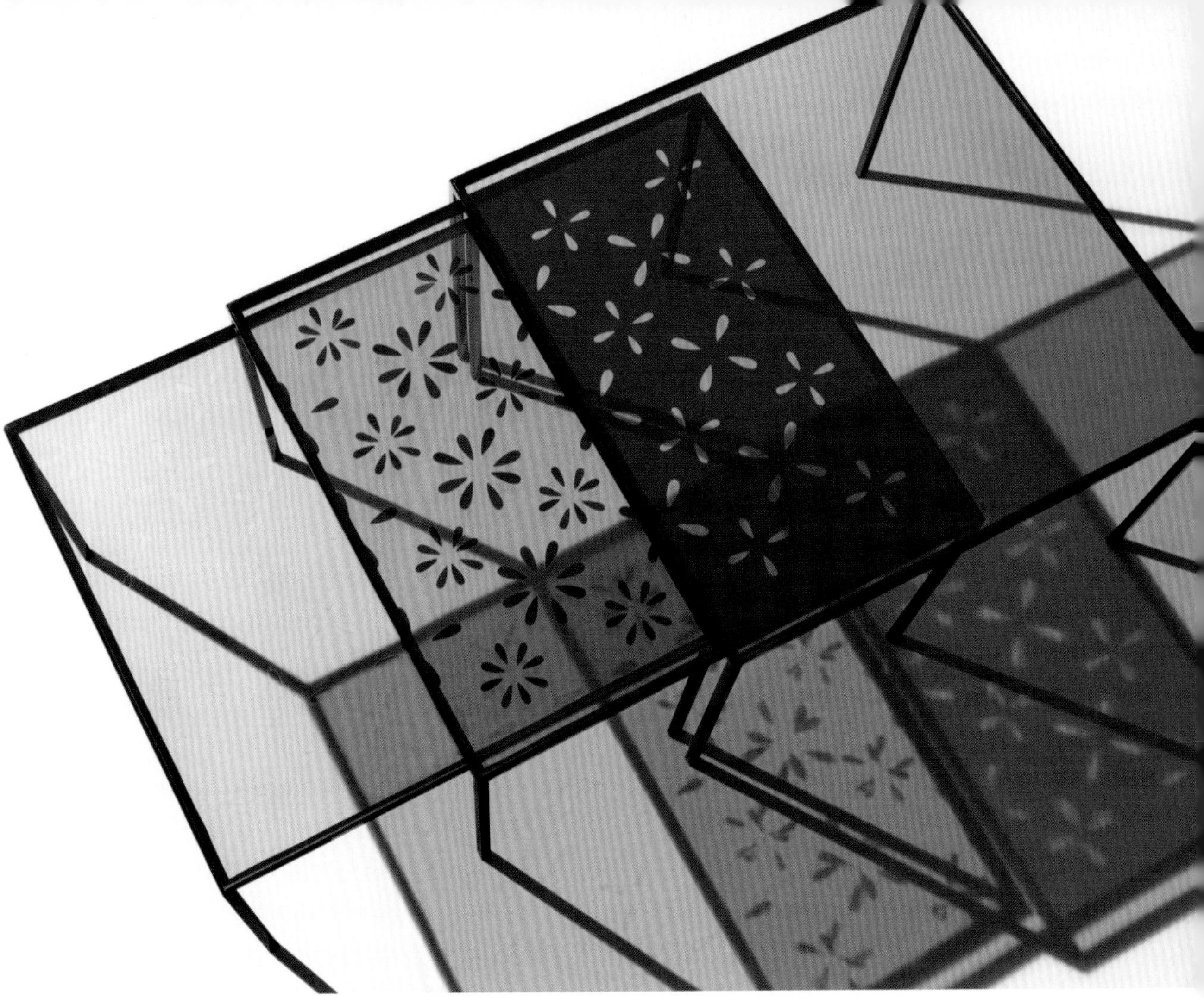

Alice's Tea Party

Lipton

An interactive cafe design based on the Mad Hatter's Tea Party from Lewis Carroll's 1865 children's classic *Alice's Adventures in Wonderland*. In celebration of the 100th anniversary of English tea brand Lipton's presence in Japan, the event featured a table and two rows of chairs that had been radically distorted in size, in order to heighten the room's sense of receding depth. Some of the chairs had been shrunk so that visitors could barely squeeze into them, while others were enlarged until the visitors' feet couldn't touch the ground. The same disorientating approach was applied to the wallpaper, which was adorned with silhouettes of characters from the novel. These details resulted in a space that felt long and narrow, as though it were physically pulling people in. The dramatic change in scale was inspired by the moment in the novel when Alice sampled the cake labelled 'EAT ME'.

Talking

H Concept

Instead of relying on the customary symbols and letters normally used on jars of seasoning, the Talking sauce and condiment containers were each designed with a mouth-shaped detail that visually expresses the pronunciation of the content's name in Japanese: 'yu' for *shoyu* (soy), 'shi' for *shio* (salt) and 'ko' for *kosho* (pepper). In an age of digital communication, when conversation tends to fade from the dinner table, these visual references to verbal communication subtly link people through the power of implied spoken language.

Chocolate-Pencils

Hironobu Tsujiguchi

Top Japanese chocolatier Hironobu Tsujiguchi invented a dessert based on his impression of nendo and, in exchange, the practice designed for him a plate and a set of chocolate pencils. The pencils come in a number of cocoa blends, varying in intensity, and can be grated onto the chocolate dessert using the accompanying pencil sharpener, adding a final design flourish. Pencil filings are usually discarded – but here, they are the main attraction.

Naoshima Standard 2

Benesse Art Site Naoshima

Exhibition signage created for *Naoshima Standard 2*, held at the Benesse Art Site Naoshima, an island in Japan's Seto Inland Sea known for its contemporary art collection.

The project included the signs installed at the three ports serving Naoshima – Takamatsu, Miyaura and Okayama – as well as all those used on the island itself, and the design for a photography exhibition on its ferries. The signs needed to be inexpensive, durable, and water- and dirt-resistant, as well as stackable for easy post-exhibition storage and reusable for future shows. Based on these criteria, white traffic cones proved the ideal solution. When lit, the cones' irregularities, caused by the metal moulds used to make them, diffused the light, giving the impression that the signage was made from Japanese handmade paper. A total of 625 cones were placed at the ferry dock in Miyaura, the islands' gateway. Once an hour, as if triggered by the waves, they were programmed to light up and welcome visitors.

The cones were also intended to act as subtle messengers for what was to come, so each was given a hand-crafted touch: a small feature that related to the artwork on show. One sign, for the piece by photographer Sugimoto Hiroshi, had a pair of binoculars perched on it as if to say 'Look over there'. Another cone, directing visitors to work by the artist Uehara Michiyo, sprouted a cat's tail, in commemoration of her feline sculptures. Cones also served as rubbish bins, ashtrays and exhibition stands to subtly unify the identity of the event.

ひさあ
ASAHI

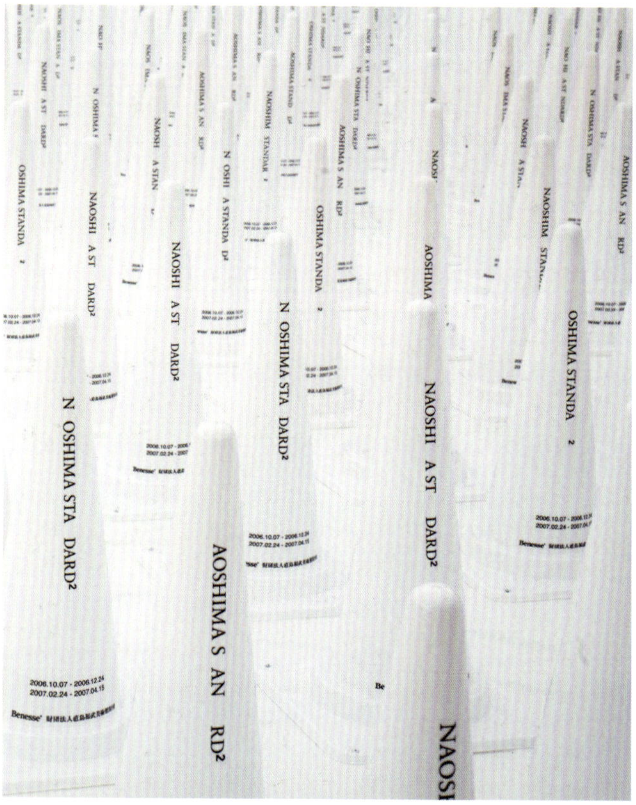
AOSHIMA S AN RD²
N OSHIMA STA DARD²
NAOSHI A ST DARD²
OSHIMA STANDA ²
2006.10.07 - 2006.12.24
2007.02.24 - 2007.04.15

NAOSHIMA STANDARD²

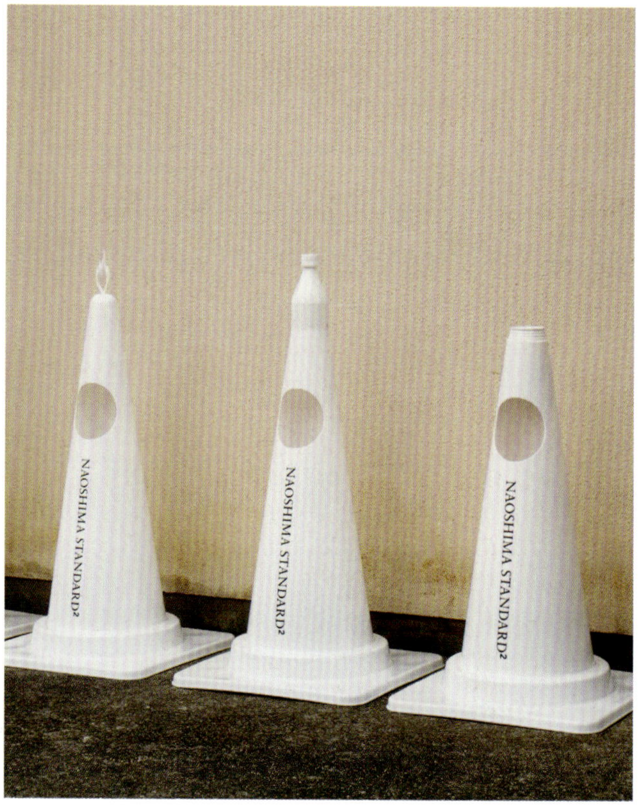
NAOSHIMA STANDARD²
NAOSHIMA STANDARD²
NAOSHIMA STANDARD²

NAOSHIMA STANDARD²
旧床屋
Former Barber

Ping-Pong Gallery
ピンポンギャラリー
NAOSHIMA STANDARD²

Bulb Lamp

1% Products

A series of distinctive lighting fixtures that give new 'personalities' to the ordinary light bulb. In one case, the twisting motion of screwing in a bulb is etched onto the form of the bulb itself. Another boasts an apparently woven surface, while the partially doubled-up outline of a third type makes it seem to sway in its socket. The final design emits light in gradations according to the varying thicknesses of its surface.

Top-Gear
1% Products

A standing bottle opener that serves its purpose no matter which way it is pushed or pulled. The hand movement used to open a bottle with it resembles that used to change a car's gear – and the more bottles of alcohol that the users open and consume, the more their level of cheer shifts up towards 'top gear'.

Vase-Vase

1% Products

Two flower vases stand on either side of an empty frame, composed like a single vase and its mirrored reflection. Placing a flower in only one vase, or putting a different kind of flower in each, creates a tiny blip in our perception of reality.

Open

1% Products

A magazine rack that captures the excitement of opening a parcel by freezing the moment at which it just reveals its contents. Shaped like a half-open box, the container's walls sit at just the right angle in order to support the magazines inside.

Fruit-Template

1% Products

A fruit bowl that looks more like a stencil or a children's shape-sorter than a traditional tabletop vessel. Instead of piling the fruit in a bowl, each piece can be individually placed in the template's many-sized holes.

1% Products at IFFT

International Furniture Fair Tokyo

An exhibition stand, designed to present nendo's 1% Products at the 2006 International Furniture Fair Tokyo (IFFT): each piece in the 1% collection was produced in an edition of only a hundred, so that purchasers could experience the satisfaction of owning 1 per cent of the designed product.

Since many exhibition halls have notoriously high ceilings, exhibitors will often attach spotlights to lower temporary walls and pillars, in order to ensure that each product is properly lit. In this instance, another strategy was adopted to avoid making the space feel enclosed. The floor was covered with a thin sheet of film, on which was printed a series of artificial spotlight markings and their shadows – one for each of the objects on display. As a result, each product stood out and the space felt masterfully illuminated without having sacrificed its spaciousness. Each 'shadow' was designed to emphasize the particular nature of the individual products, as a kind of intuitive, nonverbal explanation of them.

1%

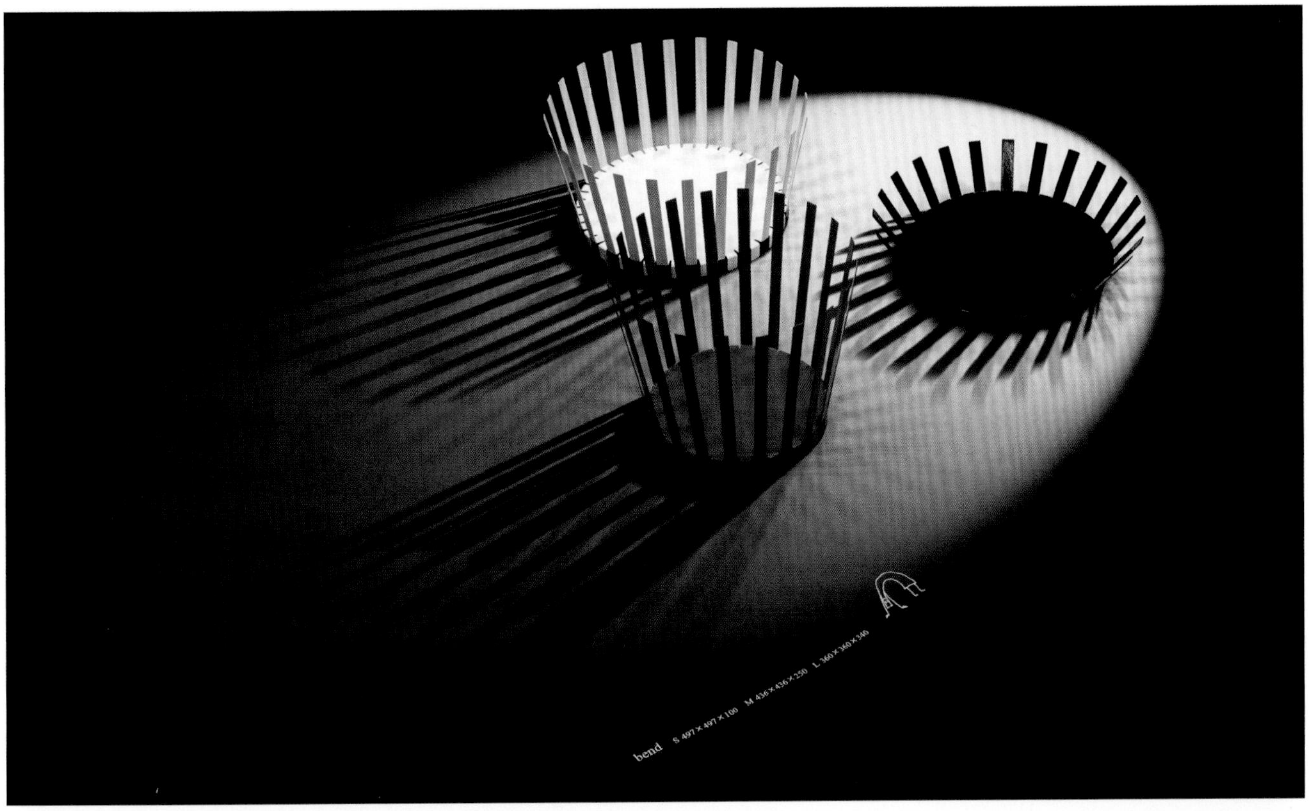
bend S 497×497×100 M 436×436×250 L 360×360×340

bend S 497×497×100 M 436×436×250 L 360×360×340

Candlever

Vogue

A candle stand consisting of a single bent-steel plate. The candle is held in place from the side using the prong provided, which means that the stand can support candles of varying heights and widths. Since the support is also a handle, and the base additionally functions to catch the melting wax, the design combines multiple functions into a minimum number of design elements. The candle stand's name is a play on the architectural concept of the cantilever, whose form it recalls.

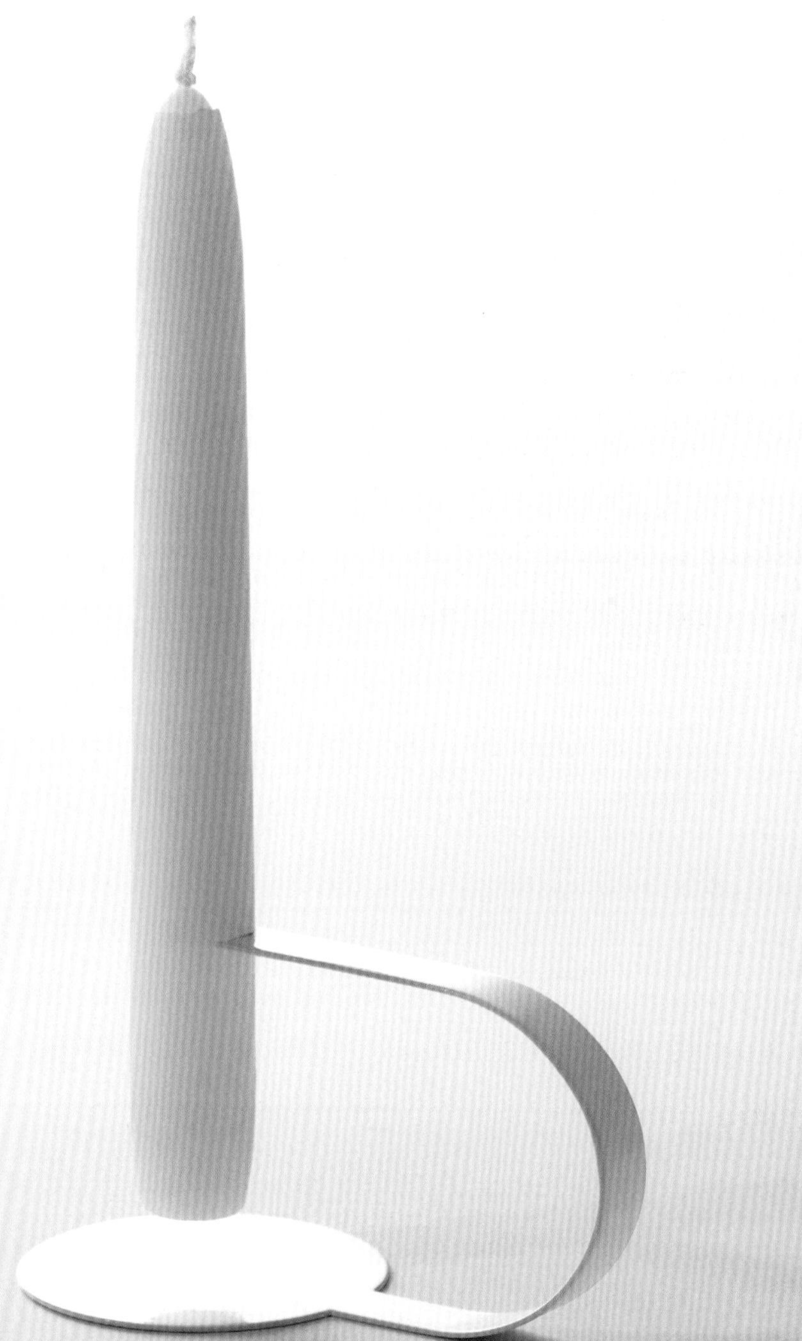

Ondle

FSB

A contribution to *Handle Couples Project*, an exhibition sponsored by architectural ironmongers FSB, which asked participants to question the assumption that pairs of door handles must match – and propose two unique handles whose differences combined to form the perfect 'couple'.

Ondle referenced the potential difference in temperaments between romantic partners. The handles share the same form, but are painted as if one is frozen while the other glows hot. The goal was to evoke three emotional states: the hesitation one might feel before touching a potentially hot or cold surface; the anticipation of opening a door without knowing what is on the other side; and the anxiety that might come from spying on your partner. 'Ondle' combines the Japanese word *ondo* (temperature) and the English 'handle'.

Illoiha 2nd Concept

Loran & Co

An interior design project for a branch of Tokyo's Illoiha fitness club, located in the upmarket shopping area of Omotesando. Since the gym was spread over two floors, a wall was constructed in order to connect the two spaces.

To express the club's original brand concept of 'becoming beautiful through movement', the theme of 'rock-climbing in Omotesando' was developed, which played on the mismatch between a rugged outdoor sport and Tokyo's eponymous fashion district. Instead of the usual rough and outdoorsy climbing wall, the new space used interior-design elements, such as picture frames, mirrors, wall-mounted deer heads, bird cages and flower vases, to create a challenging wall with hard-to-find holds and unusual finger grips.

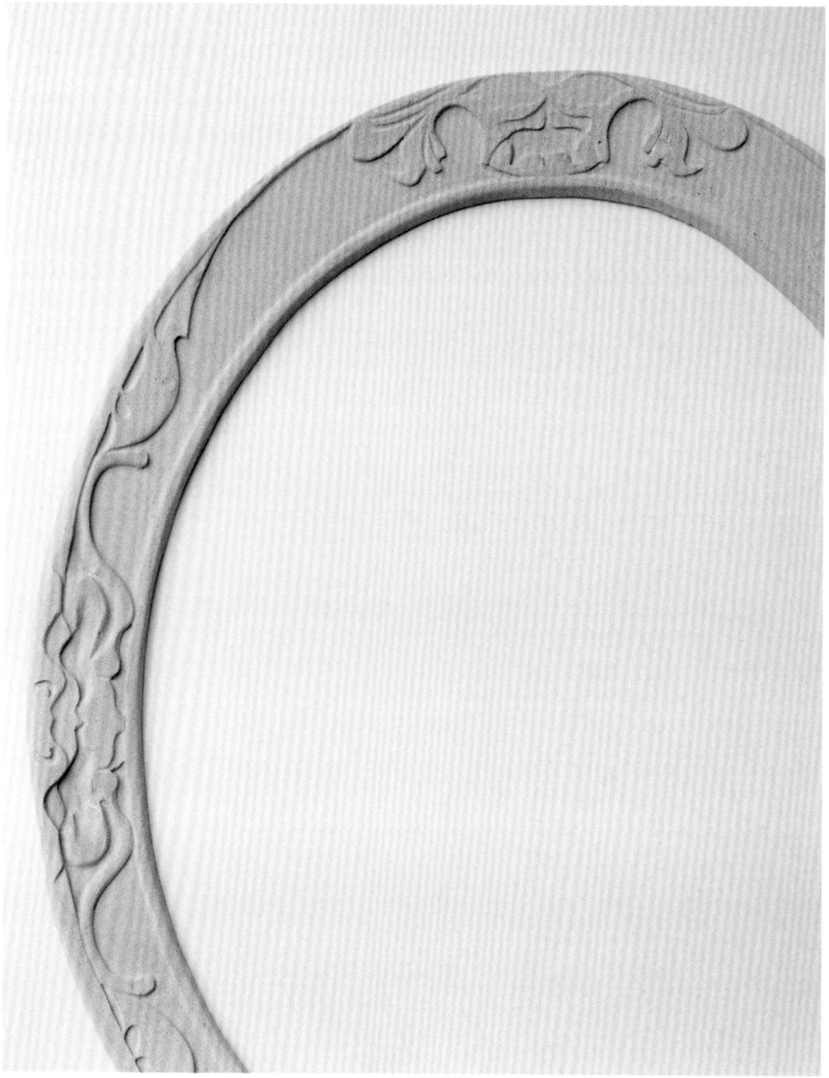

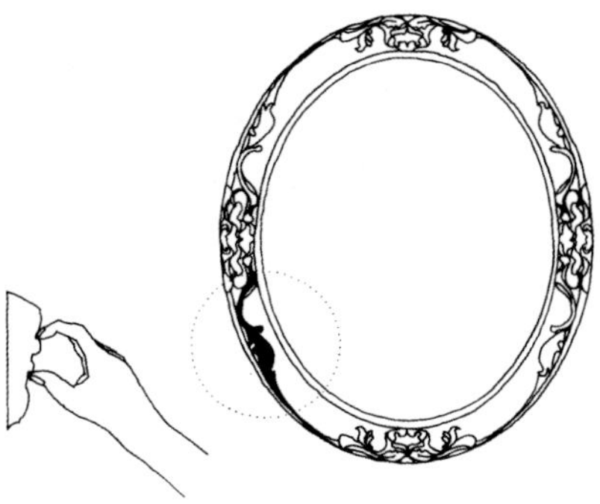

LEVEL★★★★★

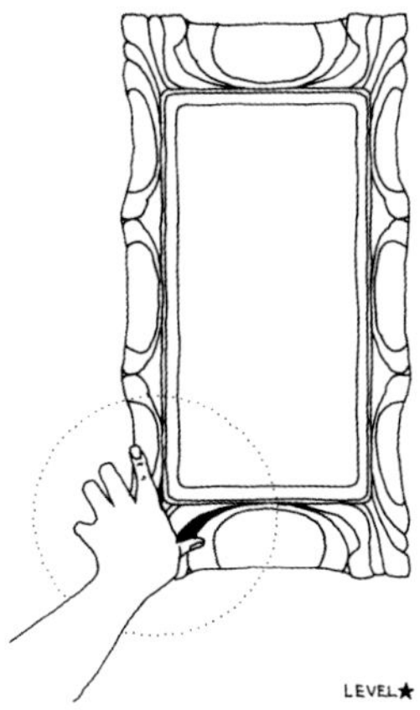

LEVEL★

LEVEL★★

LEVEL★★★

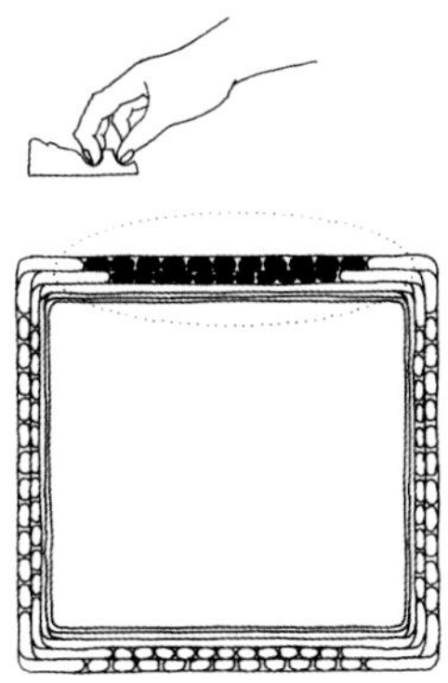

LEVEL★★★★

LEVEL★★

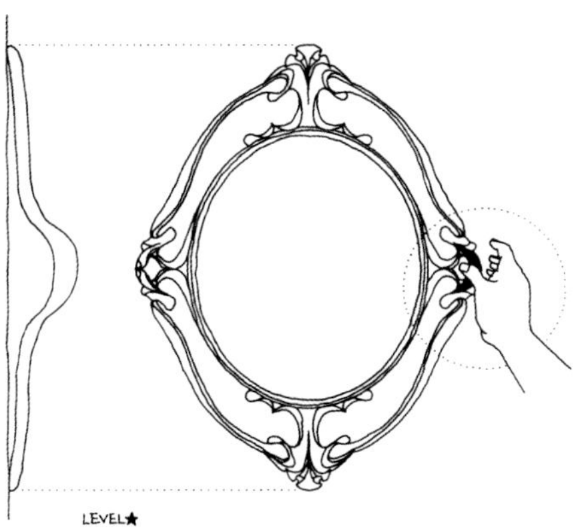

LEVEL★

LEVEL★★★

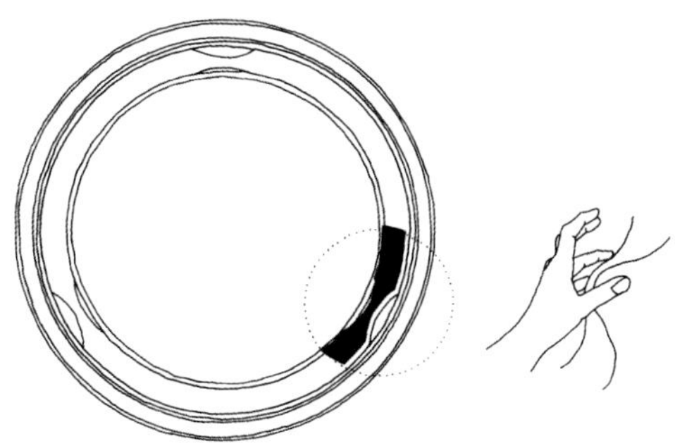

LEVEL★★

Yuki
Cappellini

A screen made of simple *yuki-* (snowflake-) shaped pieces, which connect to each other vertically in a manner reminiscent of snow piling up on the ground. Its design makes it easy to adjust the size according to the space available, and as no screws are required for the screen's assembly, it is incredibly simple to install. With each piece being the same size and shape, they can also be stacked and stored away when not in use.

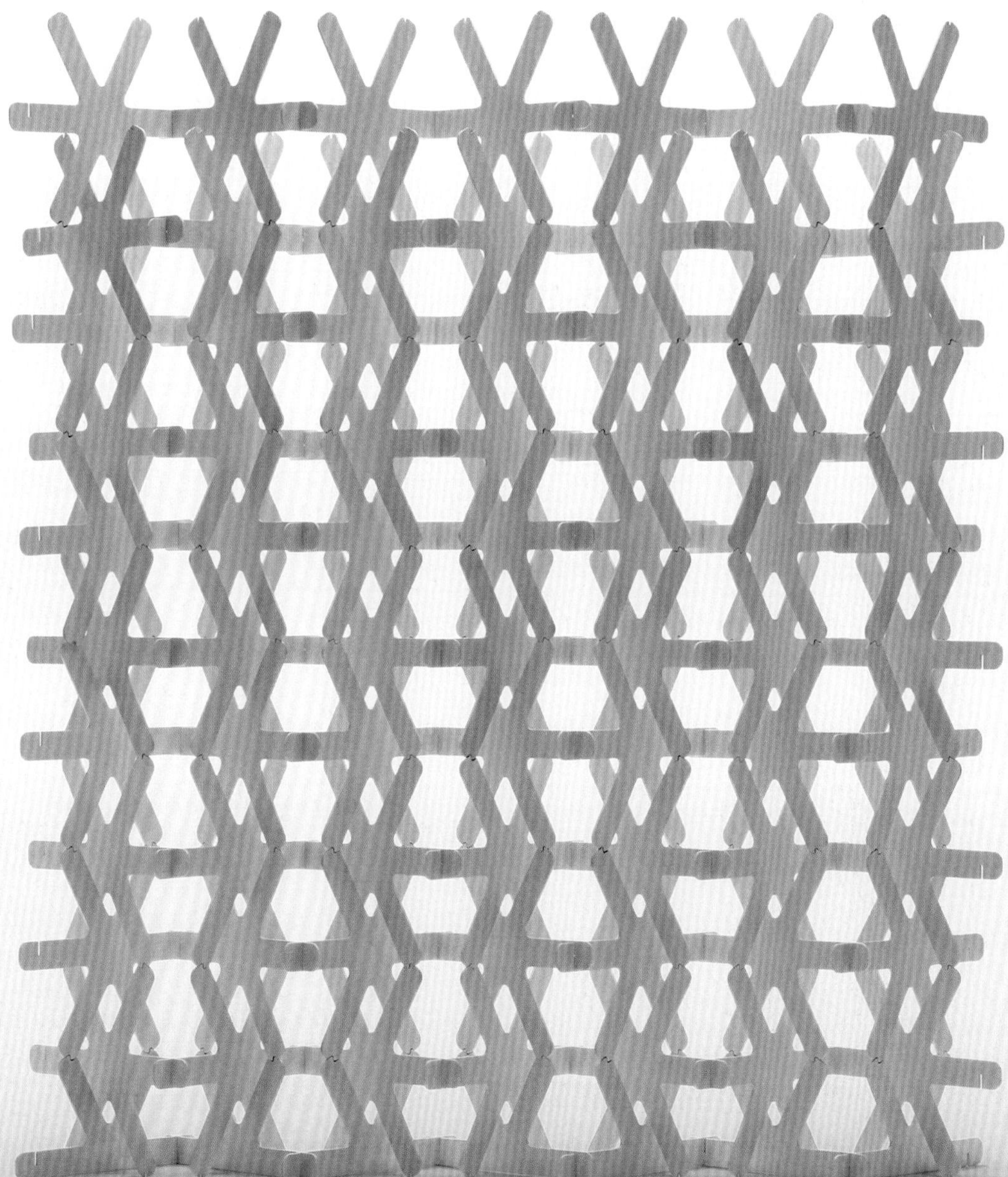

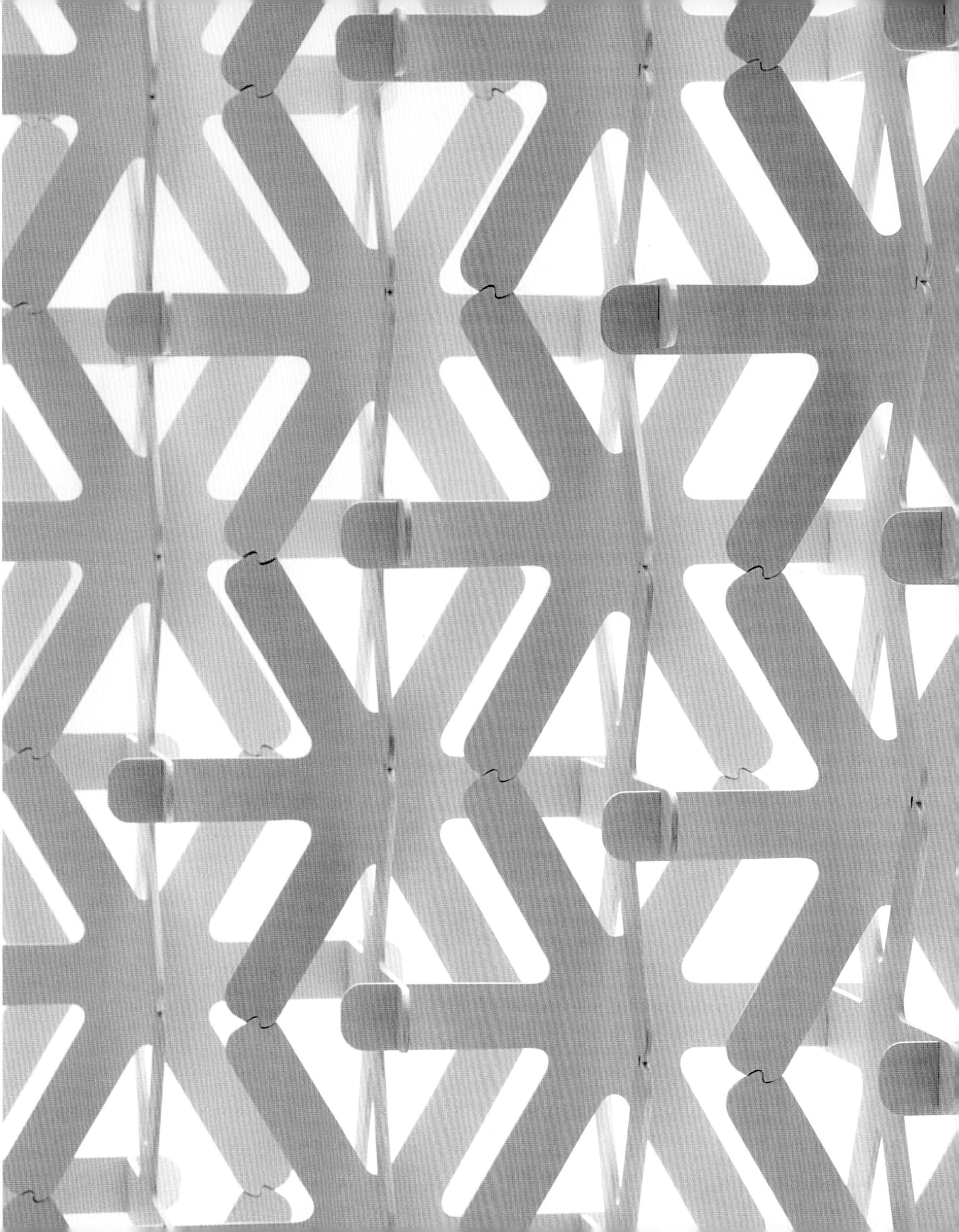

Tetris

De Padova

The two cantilevered tabletops contained within Tetris do not quite meet in the middle, giving those who notice the sense that something might be amiss. Suspended from a metal frame, the coffee table's surfaces appear to be floating, creating a distinct feeling of tension. The vision of two square planes hovering above a lower surface at the table's base – with a void of three-dimensional space in between – may bring to mind the well-known computer game that gives the piece its name.

Switch

Oluce

A floor lamp whose brightness is adjusted easily by turning its lampshade. In consolidating the light source and the dimmer switch into one fixture, the design makes users feel as though they are touching light itself.

Meguro Office

nendo

Nendo's studio between 2008–15, located near the Meguro River in Tokyo. The practice wanted the usual rooms and functions of an office environment – meeting spaces, management area, desks and storage – to be separate but also to maintain a sense of connection between them. To achieve this effect, the space was divided with walls that seem to sag and droop like a piece of cloth held up between two hands. These enclose the various spaces more than the screens and booth dividers that one might usually find in an open-plan office do, but less so than actual walls would. Employees move between spaces by stepping over the lowest parts of the walls – emphasizing, with this transition, the change in use between the different areas.

Hikidashi

Numéro

Created for the first Japanese edition of French fashion magazine *Numéro*, Hikidashi responded to a brief for a freely designed container measuring 30 x 40 x 7 centimetres (12 x 16 x 3 inches). Instead of filling an empty box with something, the idea here was to make the box appear even emptier. To achieve this, a series of successively smaller boxes were nestled inside the larger one, each with handles attached, making them resemble drawers. These 'drawers' could then be shifted around and repositioned by each user. As the drawers are arranged and rearranged, the effect is one of empty spaces appearing out of themselves, with no clear indication of whether they are being pulled out or pushed back into place. This process of constant reconfiguration was intended as a metaphor for the way in which our minds operate.

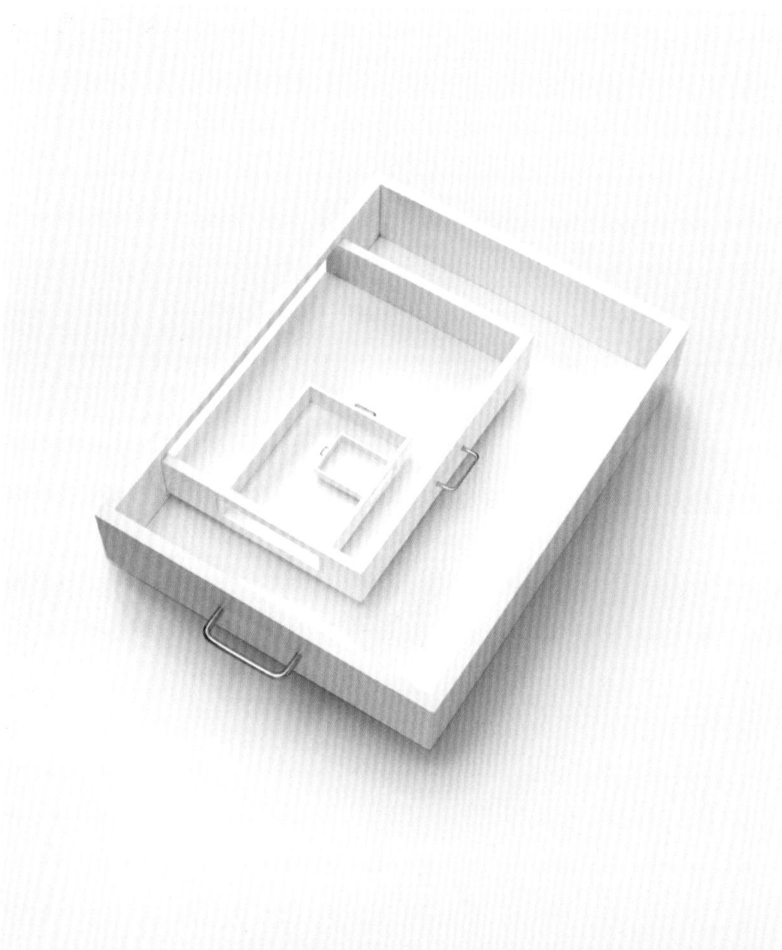

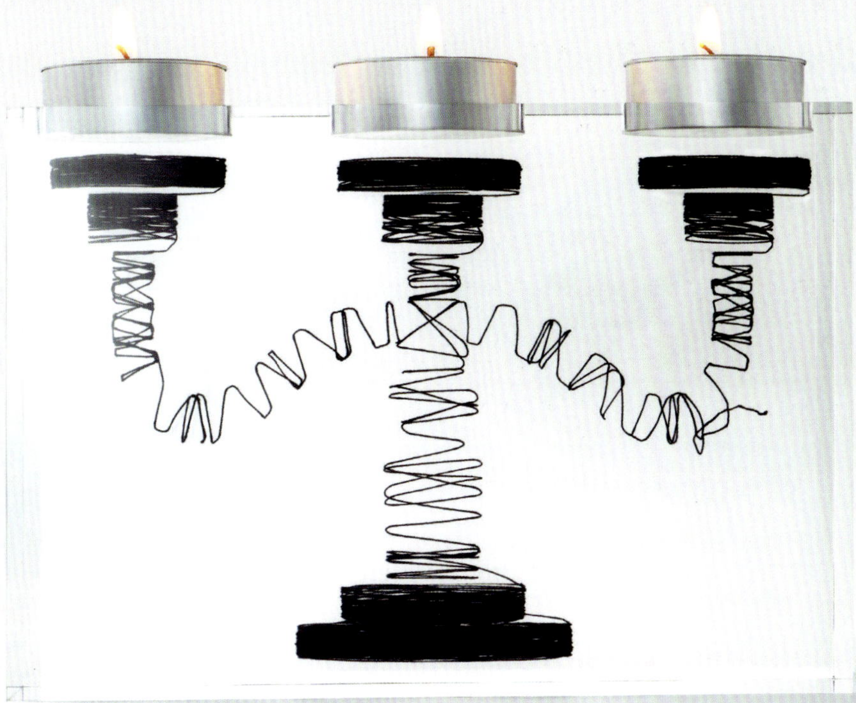

String-Candle
1% Products

A piece of string, shaped like a classic candelabra, cast inside a block of transparent acrylic, making it seem as if the stand itself is constructed from candle wick. In order to emphasize the authentic quality of the string, the design incorporated a series of deliberate irregularities that appeared to be caused through its wrapping and shaping, and sandpaper was used to scuff and soften up its surface.

Island Table

Cappellini

As the water level rises, what seems to be a solid mass of land reveals itself as a scattered collection of islands. Taking inspiration from that moment of submersion, the Island Tables branch out into two or three distinct but connected surfaces. Used together, their components can be combined in multiple ways to form a variety of 'archipelagos'.

Ribbon

Cappellini

A stool whose construction loops together like the ribbons of ballet shoes. The three 'ribbons' are laser-cut from a single, thin layer of steel, and cross at evenly spaced intervals for stability and structural integrity. The high-stool variant incorporates an extra loop at its base, which can be used as a footrest, and the optional cushion affixes to the stool with magnets.

1% Products at IL

Interior Lifestyle Tokyo

An exhibition stand design for the Interior Lifestyle trade fair in Tokyo, which was constructed in an atrium that gave visitors a view from above. In the installation, an entire living space – from sofas and tables to crockery and plants – was displayed on one wall, so that viewers looking over the edge of an upper-floor gallery felt as though they were on the ground floor and viewers standing on the actual ground floor felt as though they were looking down from above. Visitors were essentially 'walking on the walls', which heightened the topsy-turvy effect, as well as making them part of the display.

2008

Cabbage Chair

21_21 Design Sight, Tokyo

A chair designed for *XXIst Century Man*, an exhibition curated by Issey Miyake to commemorate the first anniversary of the 21_21 Design Sight museum in Roppongi, Tokyo.

Miyake challenged nendo to create a piece of furniture out of the paper that is often discarded as an unwanted by-product during the mass production of pleated fabric used in the fashion industry. The practice's solution was to transform a roll of this waste material into a small chair that is formed as its outside layers are peeled away, one by one.

Resins added during the original paper-production process lend the material strength and the ability to 'remember' forms. The pleats themselves give the chair elasticity and a springy resilience, for an overall effect that looks somewhat rough but that gives the user a soft, comfortable seating experience.

Since the production process was so simple, the idea was that eventually the item could be shipped as a single compact roll for the user to cut open and peel back at home. The chair has no internal structure; it is assembled – or perhaps more appropriately, given the peeling process, 'disassembled' – without nails or screws. This primitive design approach confronts the problems that society must address in the new millennium, such as manufacturing and logistics costs, as well as environmental issues. Thus, the Cabbage Chair fits active, optimistic and forward-thinking 'twenty-first-century people' – the kind who, to borrow a concept that Miyake himself expressed, 'shed their skin' in the drive for self-betterment.

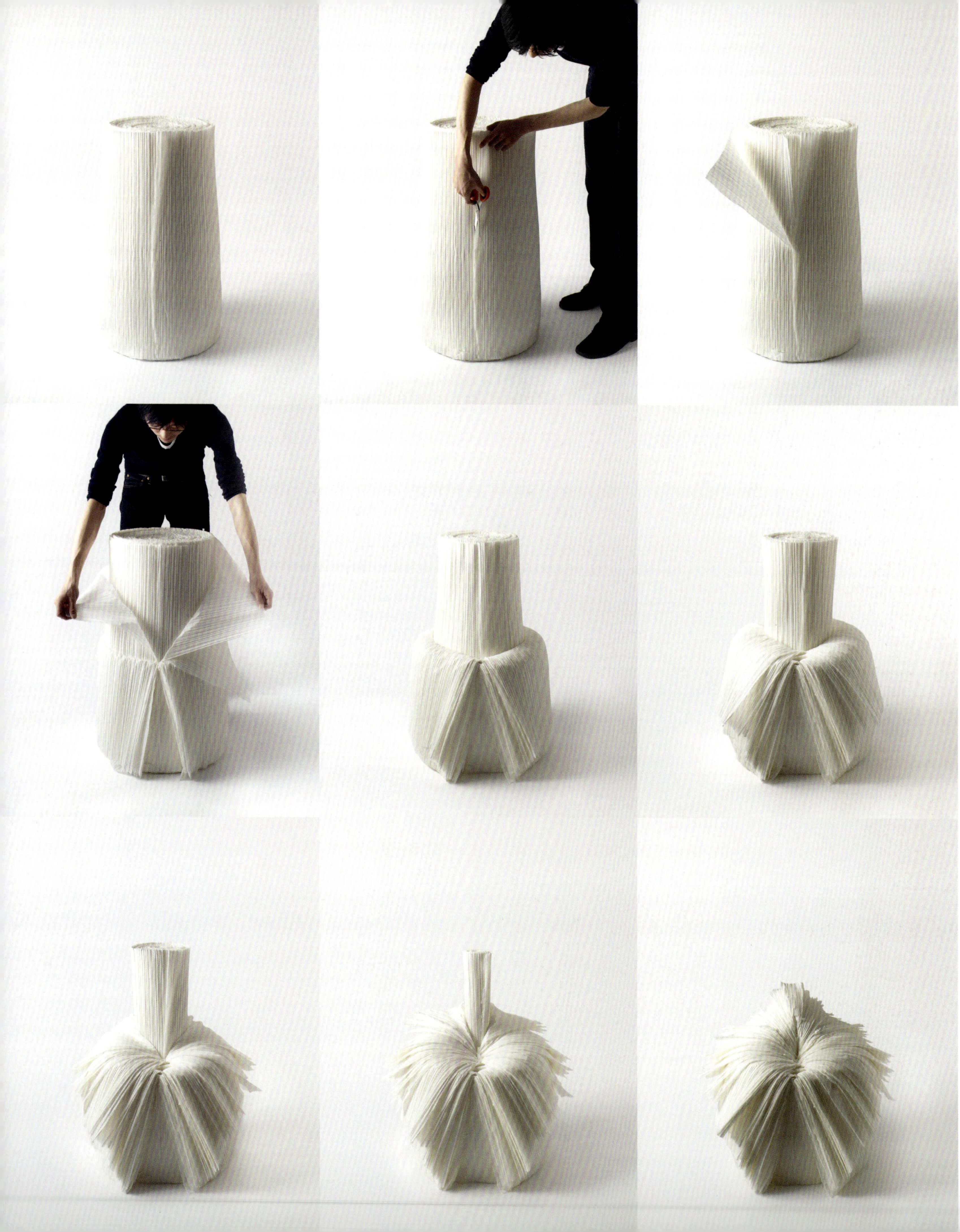

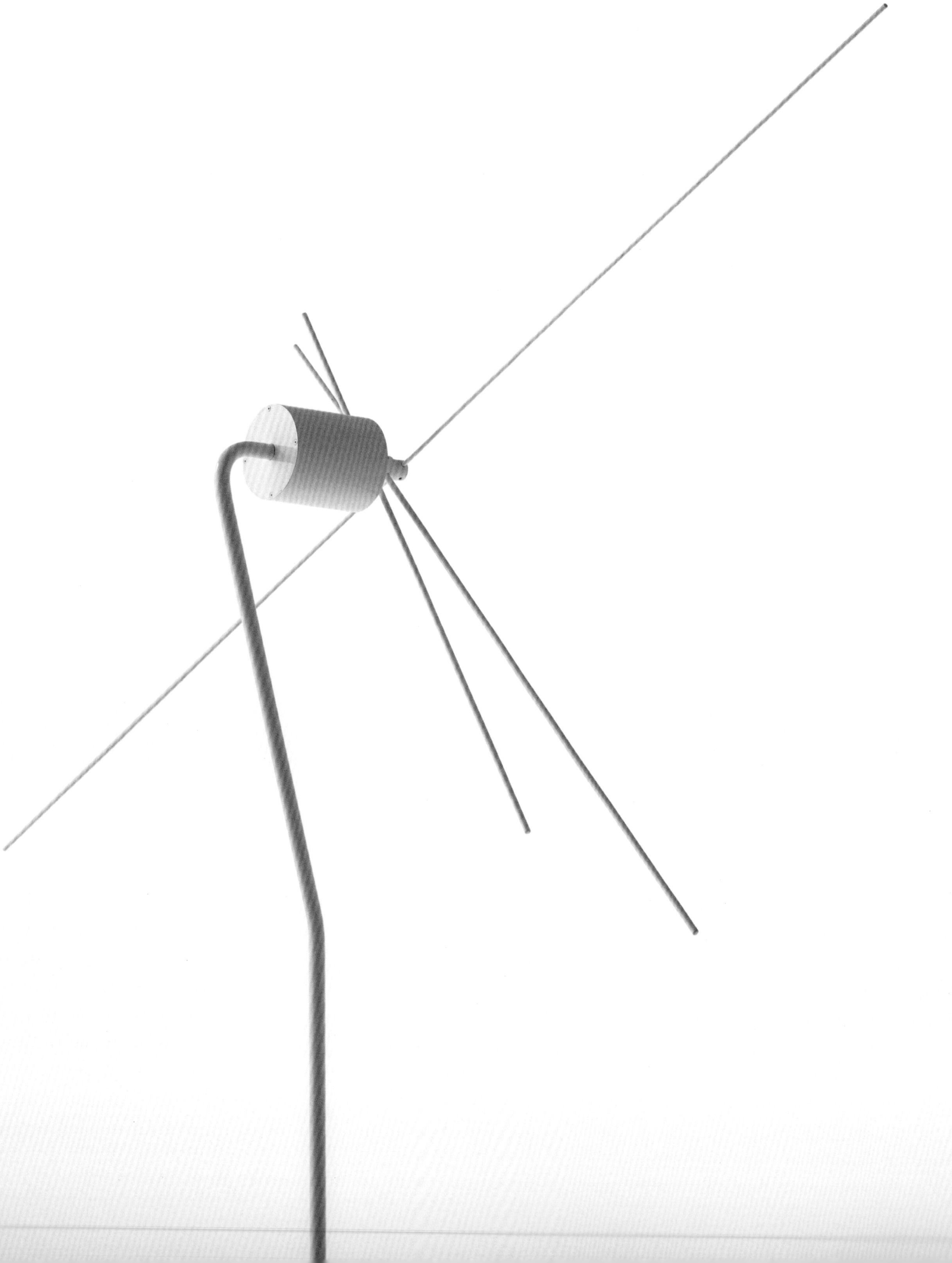

Kazadokei

Museum of Contemporary Art, Tokyo

A tall and narrow floor-standing clock, measuring 2 metres (6½ feet) in height, and with a second hand that is 150 centimetres (60 inches) long. The design's proportions require it to use the same kind of mechanism employed in large, public timepieces for buildings and parks. When several Kazadokei are lined up side-by-side, they resemble a series of windmills – this visual similarity thus forming the inspiration for the name, *kaza-* (wind-) *dokei* (clock).

Today, in an era in which our perception of time has changed quite significantly, Kazadokei lets us experience it directly with our bodies and senses. Similarly to windmills, which revolve when they catch the wind, Kazadokei revolves as it 'catches' time. While watching the time pass, one might even feel a gentle breeze as the 'blades' rotate through their continuous cycle.

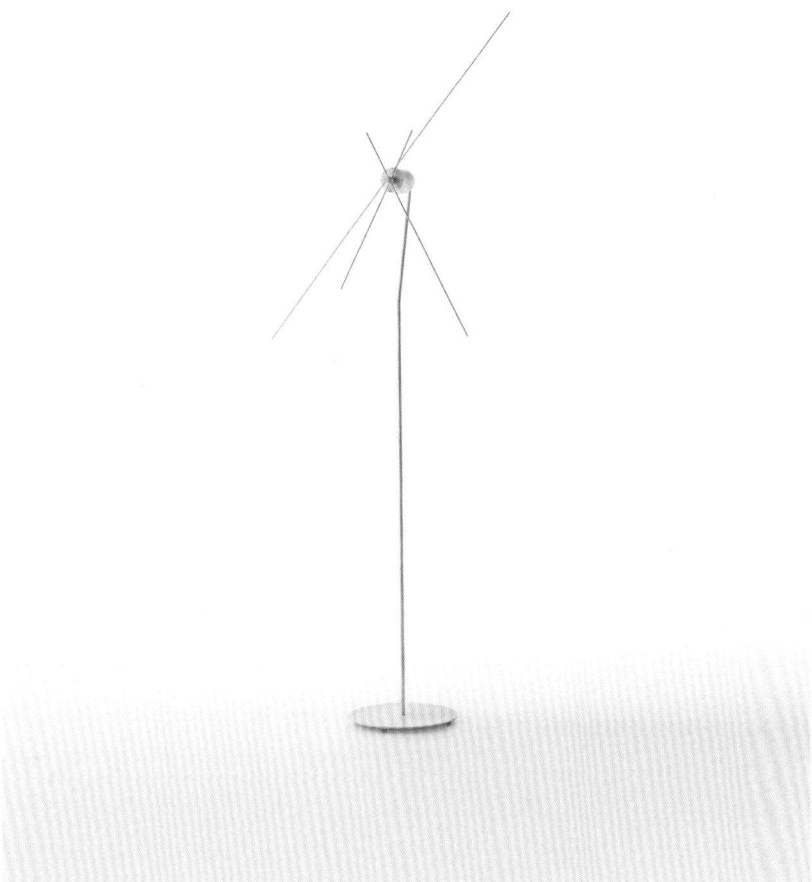

Moss House

Private Commission

A renovation project of an old wooden house on the Shibuya River in Tokyo's Ebisu neighbourhood. Moss patterns the riverbank outside the property's windows, so similar veins of dry moss were used on its internal walls to subtly connect the house's interior environment with its exterior one. Most wallpaper patterns imitate nature through a two-dimensional representation of it, but cladding the walls entirely in moss would have been excessive. Something 'in between' was required, so the moss was applied in a pattern that looked like that of wallpaper motifs, creating an ambiguous texture that is neither artificial nor natural. The same pattern appears on the cable outlets, door frames and handles, further uniting the moss pattern with the space.

Ghost Stories

Friedman Benda, New York

An installation of Cabbage Chairs in New York's Friedman Benda Gallery. To heighten the sense of drama created by the top-light pouring from the rear of the gallery, and to emphasize the chairs' layered gradations of light and shade, 6,800 threads were suspended from the ceiling. Their shadows made the space look as though it had itself been pleated. Additionally, by colouring each thread at different heights with permanent black marker, a 'boundary between light and shadow' was created in an attempt to emulate the natural light entering the gallery.

Inspired by the shifting appearance brought about by these effects, and by the way that the edges of objects seemed to blur, the installation was named *Ghost Stories*.

Socket-Deer

Au

Electrical-outlet covers that allow users to hang their mobile phone on the wall as it recharges. The three types of deer antlers were chosen for their natural ability to hold things, so there was little need to modify their orginal forms. The tough urethane rubber used for the covers holds the handsets tightly, and also protects the antlers from breakage should a passer-by accidentally bump into them. Socket-Deer can also be used as a cover for light switches, and the antlers make an excellent hook for keys or other accessories.

Diamond Chair

Lexus

A diamond's dense atomic structure efficiently disperses strength and light, giving the gem its trademark hardness and shine. This 'strong but flexible' composition inspired the creation of a chair that would respond to pressure by absorbing rather than resisting it, and that had the suppleness of human muscles. Powder-sintering rapid prototyping (RP) technology was used to add thickness where users need support and to carve away material to a thinner, more flexible and responsive depth where they want comfort – one material thus serving multiple purposes. As there was a limit to how big an object the RP machine could produce, the chair was designed in two pieces that snap together easily after each has hardened. This not only reduced costs and production time but also, since the pieces are cast simultaneously, eliminated the possibility of mismatching parts during assembly. Any shrinkage or heat distortion occurs to the same extent for both pieces in each pair.

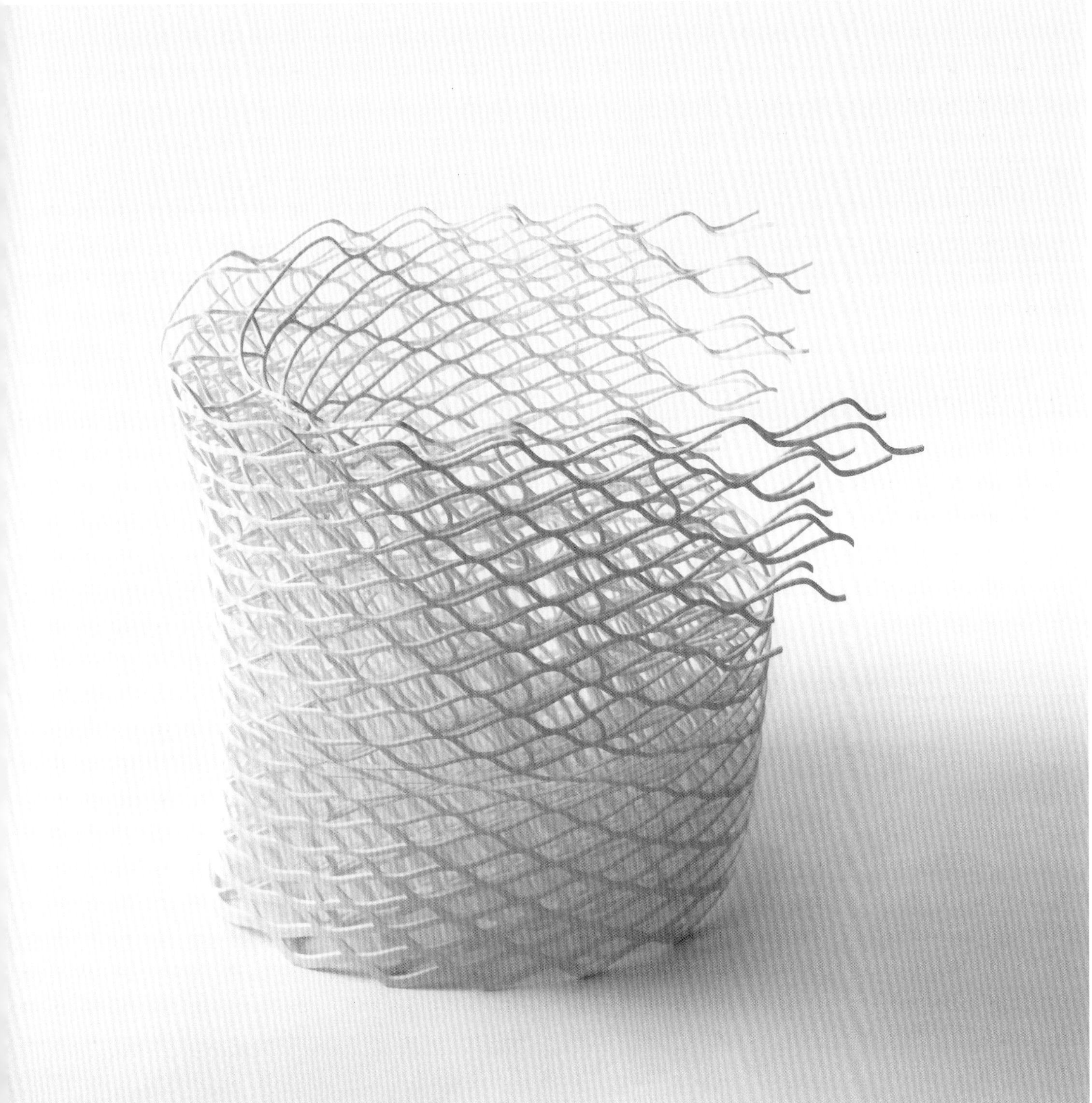

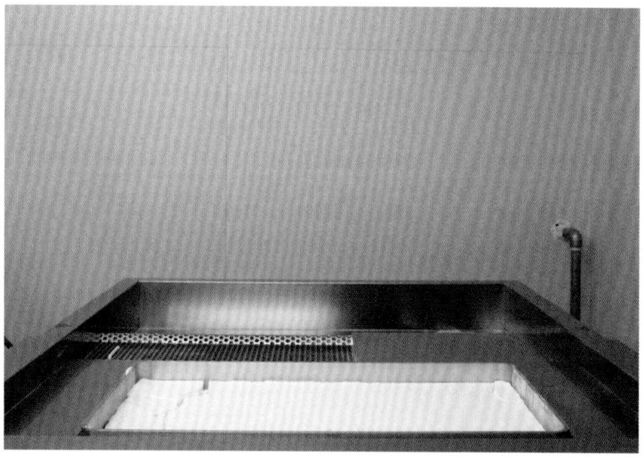

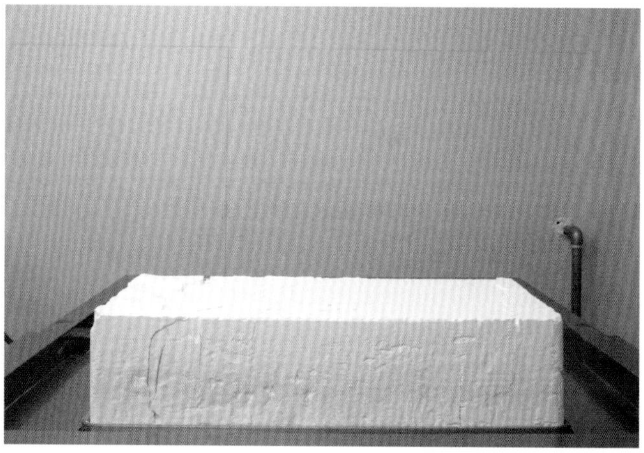

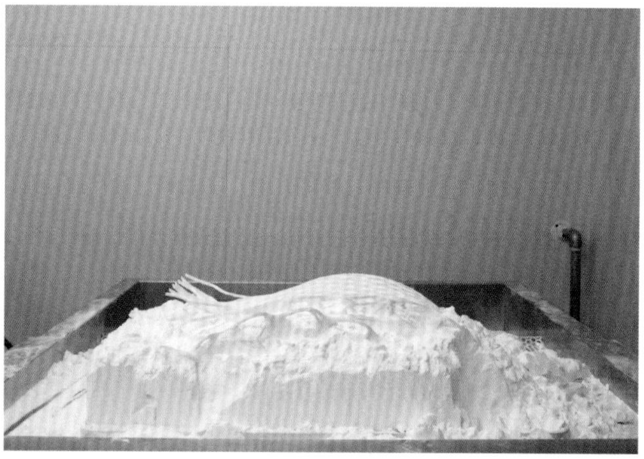

Lexus -Elastic Diamond-

Milan Design Week

An exhibition at the Museo della Permanente in 2008, during Milan Design Week. The design allowed visitors to experience the ultimate fusion of opposite elements that is 'L-finesse', the design ideal behind the premium Lexus automobile brand.

Using the crystalline configuration of diamonds as a motif, a 'flexible but durable' pillar structure was produced that fused the opposites durable/strong and flexible/fragile. This construction was tangible but permeable; it seemed complex but also natural and uncontrived – expanding and contracting as though breathing. The pillars 'shivered' in response to human movement, creating a spatial experience in which people and objects were brought into sync.

Rokumaru
De Padova

Each of the six hooks on the Rokumaru coat stand is placed 60 degrees apart from the last so that their ends meet perfectly in the centre. Shaped like two equilateral triangles, the hooks are also oriented at 60 degrees from the vertical and each one is sliced in cross section, allowing items of clothing to be hung comfortably from their ends. Appropriately, Rokumaru combines the Japanese words *roku* (six) and *maru* (zero; literally, 'circle') – in other words, 'sixty'.

Bambi
Cappellini

A desktop comprising a single plane of laser-cut aluminium, filled with polyurethane foam for an ultra lightweight structure. Each of the four legs is attached to a different edge of the table top, and therefore folds in a different direction. This gives the desk a sense of unsteadiness at first glance that is reminiscent of the well-known Disney character Bambi, the baby deer who initially struggles to stand. However, the user soon realizes that it is actually a strong, well-balanced form.

Hakuhodo Creative Vox Office

Hakuhodo Creative Vox

An interior design project for an office space that is constructed around a central terrace. The intention behind the design was to keep any new walls and furnishings to a minimum in order to preserve the airy, comfortable feel that the space already possessed. However, one of the major questions was how to deal with the ample light that poured in through the large windows. Blinds were the obvious, most functional answer, but there was a risk that this would make the space feel too formal. Accordingly, new blinds were cut in the shape of loose, horizontal waves for a gentle, draping three-dimensionality. The resulting effect, halfway between blinds and curtains, not only controls light and sight lines but also gives the entire space a more relaxed feel.

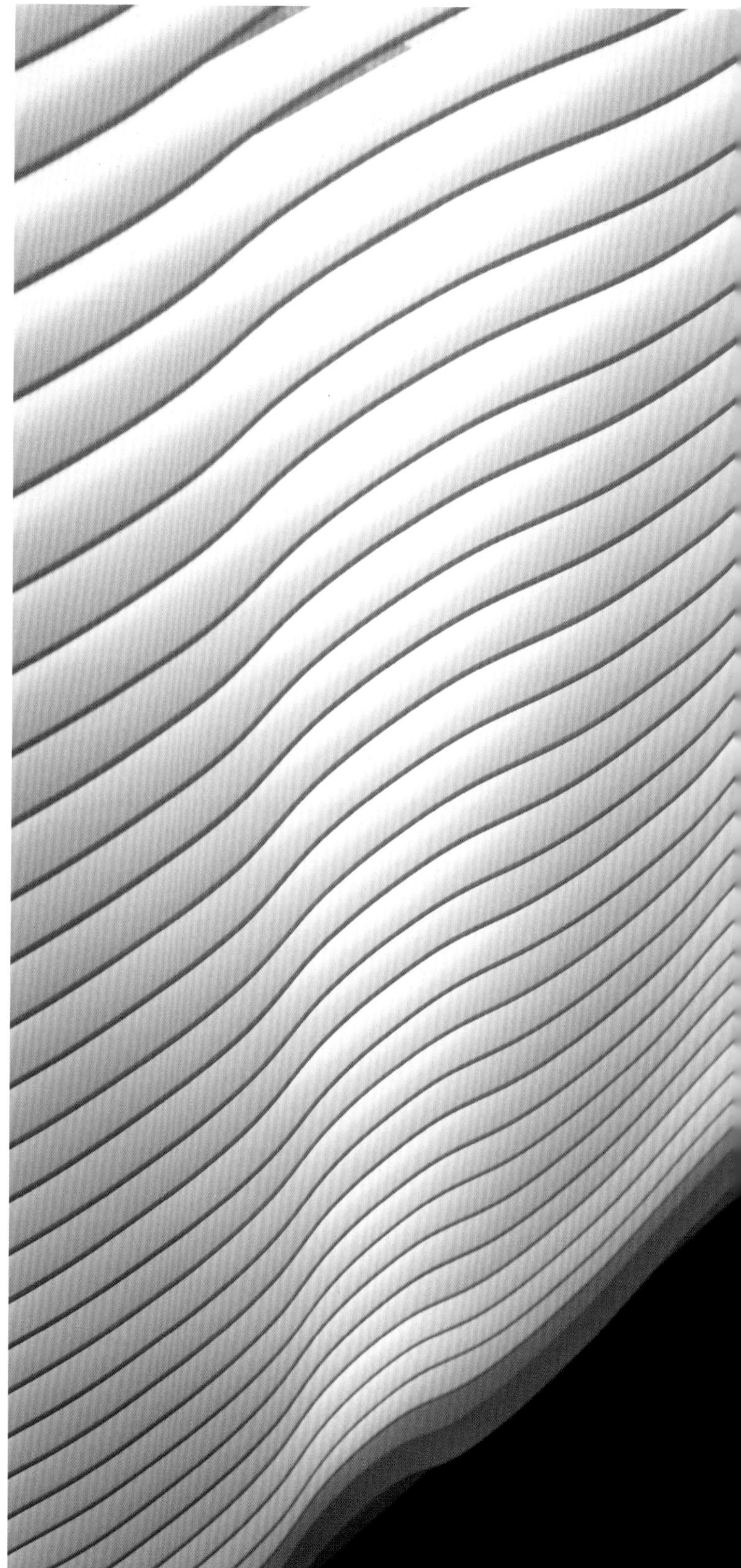

Kuuki

Kenzo

An installation that captured the concepts behind Kenzoki, a skin-care range from Kenzo that pleases all five senses using the natural elements found in plants. Plants manage water content through absorption and transpiration, their fragrance attracts pollinating insects, and they purify the air via photosynthesis. Each has only a tiny effect on its own but a group of plants creates a wonderfully pleasant environment.

For the project, six different products were designed following the attributes of particular plants, and were endowed with the same kind of near-imperceptible functions as those described above. The goal was to create forms that, when grouped, would turn space itself into a singular product, creating a captivating and charming environment.

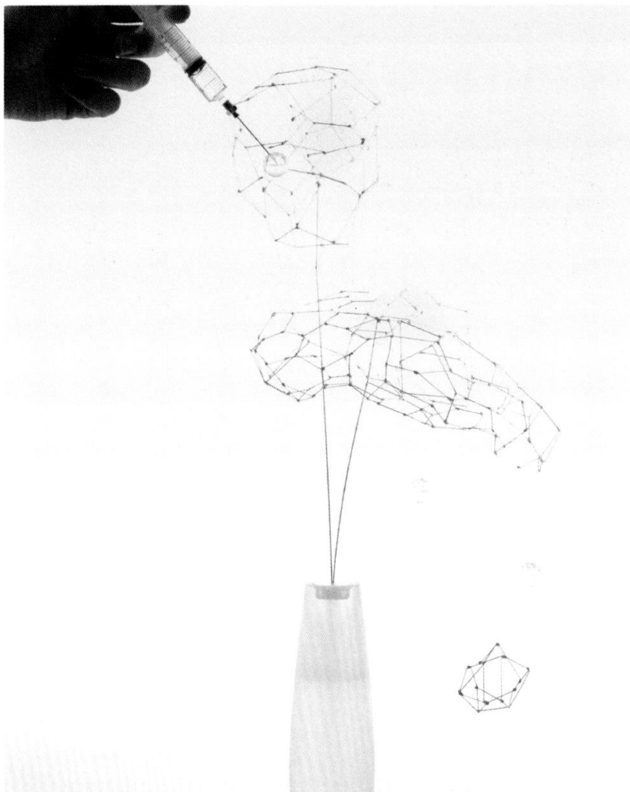

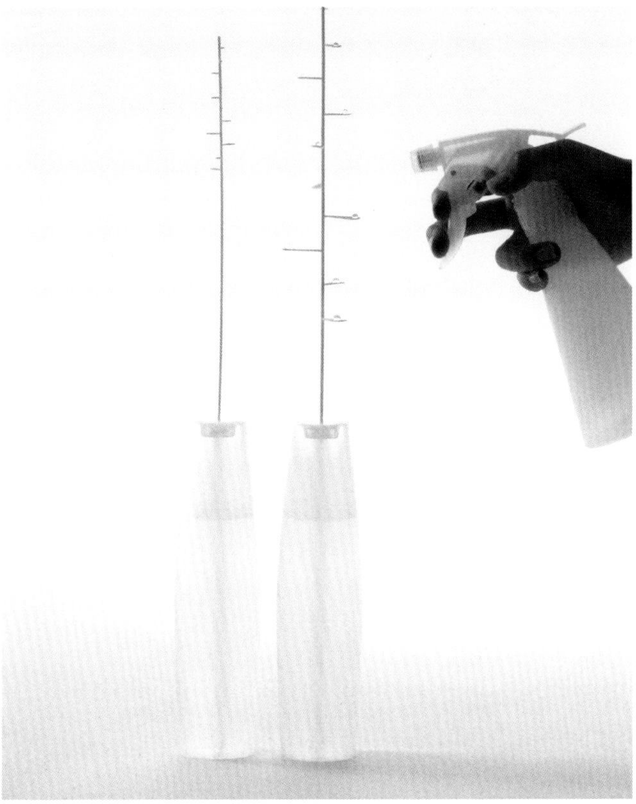

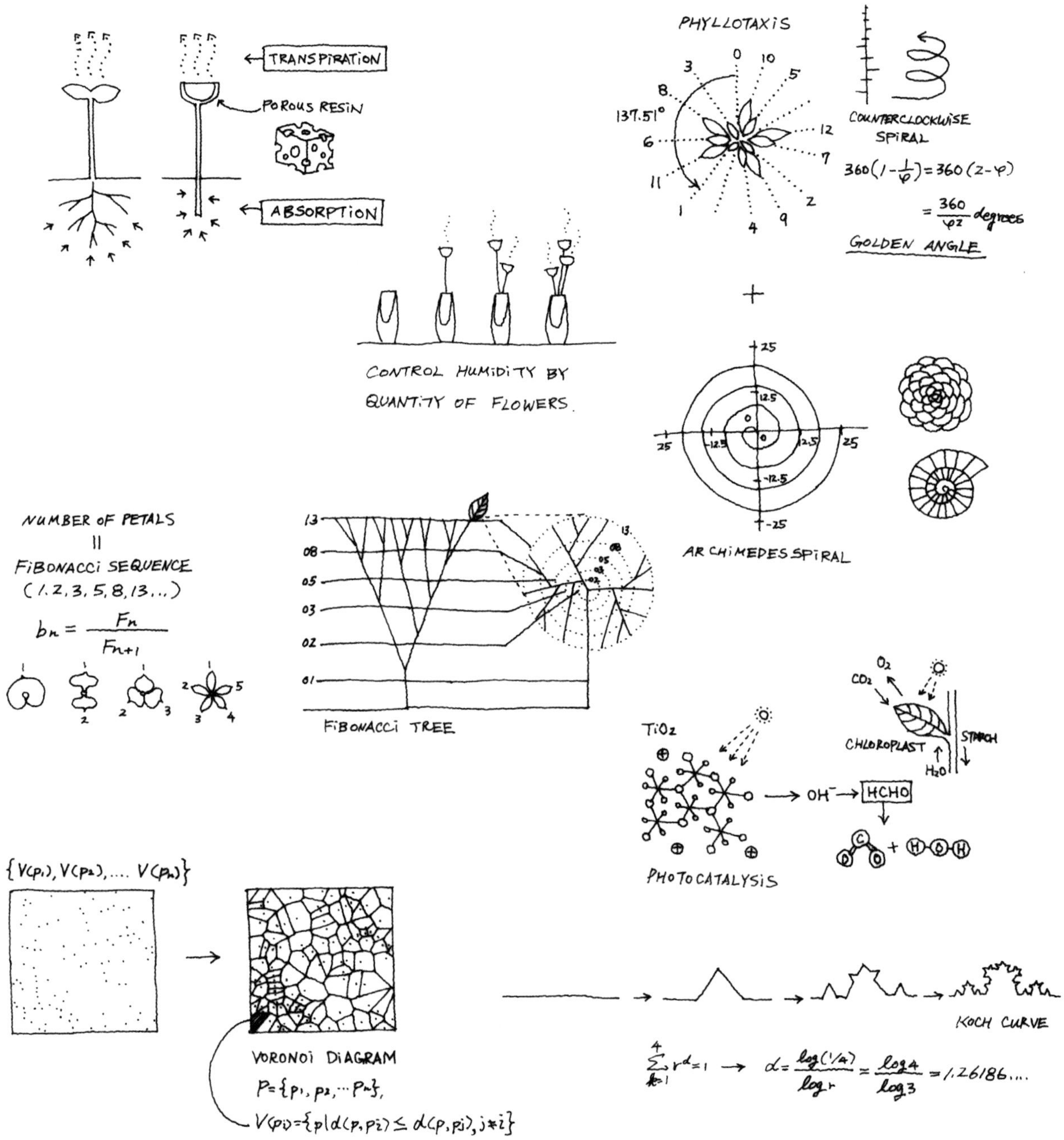

TRANSPIRATION
POROUS RESIN
ABSORPTION
CONTROL HUMIDITY BY QUANTITY OF FLOWERS
PHYLLOTAXIS
137.51°
COUNTERCLOCKWISE SPIRAL
360(1-1/φ)=360(2-φ)
=360/φ² degrees
GOLDEN ANGLE
ARCHIMEDES SPIRAL
NUMBER OF PETALS
||
FIBONACCI SEQUENCE
(1, 2, 3, 5, 8, 13, ...)
bn = Fn / Fn+1
FIBONACCI TREE
TiO2
PHOTOCATALYSIS
OH⁻ → HCHO
CHLOROPLAST
CO2
O2
H2O
STARCH
{V(p1), V(p2), ... V(pn)}
VORONOI DIAGRAM
P={p1, p2, ... pn},
V(pi)={p|d(p,pi) ≤ d(p,pj), j≠i}
KOCH CURVE
Σ r^d = 1 → d = log(1/4)/log r = log4/log3 = 1.26186....

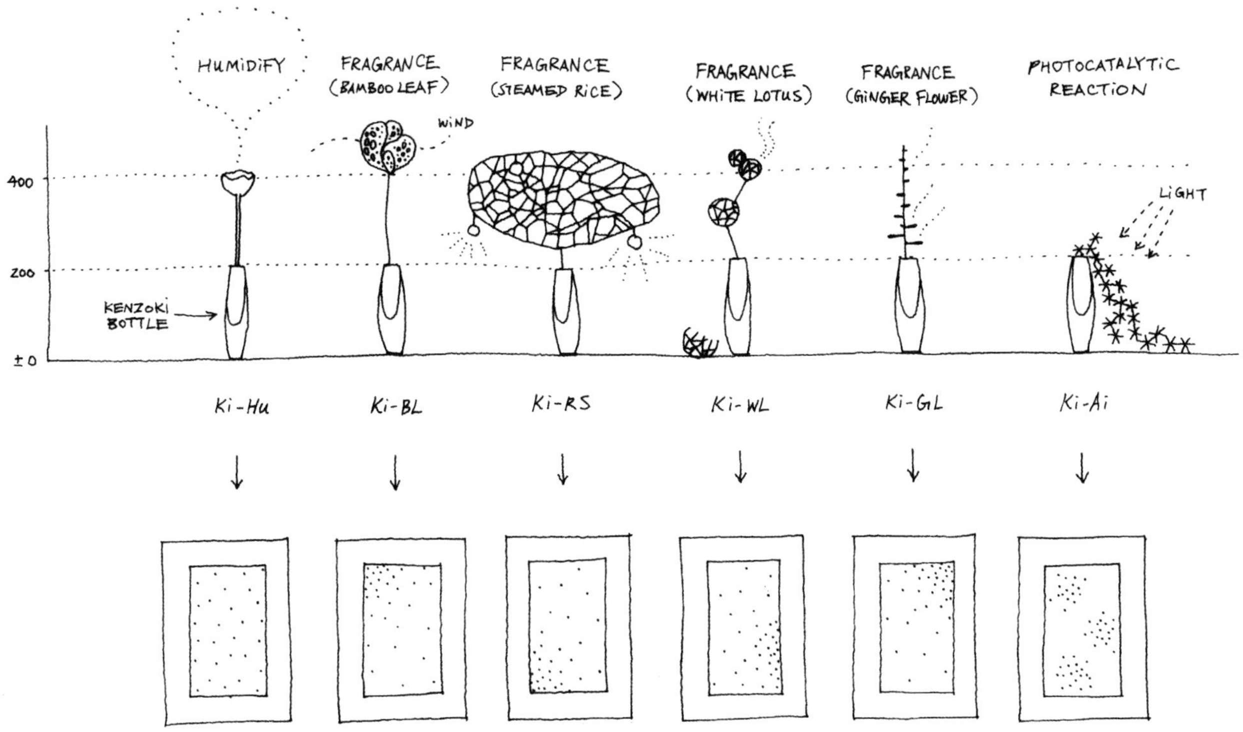
HUMIDIFY
FRAGRANCE
(BAMBOO LEAF)
WIND
FRAGRANCE
(STEAMED RICE)
FRAGRANCE
(WHITE LOTUS)
FRAGRANCE
(GINGER FLOWER)
PHOTOCATALYTIC
REACTION
LIGHT
400
200
±0
KENZOKI
BOTTLE
Ki-Hu
Ki-BL
Ki-RS
Ki-WL
Ki-GL
Ki-Ai

Catalogue of Works

2002–5

Al(Pb)_Lumi
Tokyo Designers Block

An ultra-thin, hanging lighting fixture that uses an organic electroluminescent device as its light source. The 5-millimetre (1/5-inch) -thick, ring-shaped form is made from a combination of aluminium (Al) and lead (Pb) – materials whose differing weights are counterbalanced to enable the lamp to stay level, similar to a mobile.

Cone-Light
Milan Design Week

A cone-shaped lamp shade made of linen rope. The design emphasizes the rationale behind the shape of a traditional traffic cone, an object for outdoor use, by changing its material, and adapting it for an interior light fitting. Similar to its inspiration, Cone-Light is easily portable and can be stacked to allow for variation in the quality of light and shadows.

Closet
Closet, Toyko

A gallery located in a quiet residential district of Tokyo. Unlike the usual open-plan, white and bright exhibition spaces, Closet was instead a small, partitioned space with dark wood features. Drawing on the idea of a walk-in wardrobe, the design included framed artworks that hung from hooks like items of clothing. This feature offered customers the ease and flexibility to hang purchased works throughout their homes.

Crack-Vase
Milan Design Week

We are heartened when we see a flower growing through the cracks in an asphalt-paved road. By transforming this everyday experience into the features of a vase, Crack-Vase emphasizes the power and beauty of even small traces of nature in the urban environment.

Manhole-Mat
H Concept

A bath mat that draws on the visual and functional qualities of a manhole cover. When you see a manhole on the street, it signals the all-important task of collecting and disposing of rainwater. If, on the other hand, you see one on your bathroom floor, the indication is that it should be used for collecting and absorbing the water on your feet after taking a shower.

Kemuri
En-Rest

A restaurant design inspired by the *kemuri* (smoke) produced by a charcoal barbecue, which enhances the perceived flavour of every dish. The project featured 900,000 straws that were stacked up into columns and lit from underneath in order to create a softened, 'smoky' atmosphere. These unconventional structures served as dividers between the dining booths in the restaurant.

Time
Cibone

A dishware collection that is completely white save for the colours on the base of each piece. In use, these colours appear elusively, reflected on the tabletop. Each item in the set has a different hue, so that the colours change according to the individual courses of an entire meal – from orange to red and purple to blue, like the tints of an evening sky.

Tree-Table
Idée

A bar counter made of larch that was sourced in Nagano Prefecture. The undulating shape of the countertop, which encourages greater interaction between staff and customers, was inspired by the form of a great tree, under whose canopy various animals and insects coexist.

Tipsy's
Sigma

Upon entering Tipsy's wine bar, the wooden floor sinks slightly underfoot – an effect that was achieved through layering the floorboards on top of polyurethane foam. The floating feeling that this evokes resembles tipsiness – often considered one of the more enjoyable side effects of drinking alcohol.

Fruit Tape
New Design Paradise

Packing tape printed with a surface that mimics the colouring and texture of various fruits. When opening a packed box, the user is reminded of the feeling of peeling fruit – making this a design for unpacking rather than packing.

Sorane
Oluce

A transparent, acrylic disc that emits sound and light from above, just like distant thunder and lightning. The effect is achieved through two different technologies: a vibration speaker that is activated when the user stands directly underneath the disc; and tiny dots laser-etched in a precise arrangement on the disc's acrylic surface, in order to reflect the light being emitted from the centre of the device. Fittingly, *Sorane* means 'sound of sky' in Japanese.

Ukki
Bals

The Ukki (floating) stool uses wet-suit fabric to create a sense of buoyancy when the user sits down, as it moulds naturally to the body. For additional support, the seat is made from two layers of fabric, each with a different level of flexibility, which expand gradually according to the individual's weight.

House of Trinity
Cartier

An event space designed to mark the launch of a new collection of classic Trinity rings from Cartier. The building was transformed using stereogram printing, which blended and harmonized with the symbolic colours of the ring. While the exterior facade combined three colours – yellow, blue and pink – which changed according to the angle of view, the interior was fitted out with single-coloured stereogram prints.

Skiima
Dinos

An extended sofa from which side tables and lighting fixtures sprout between the cushions. This adds functionality to what would normally be dead space and creates a new relationship between accessories and the sofa. The result is a new type of integrated furniture design.

Spice No. 81
Spice No. 81

A new identity and interior design for a clothing and household-goods brand that shows exceptional attention to materials and production processes – an approach that could be compared with the subtle, 'hidden' flavours in Japanese food. The store design was intended to have the same effect as these refined culinary details – discernible only to the discerning. Natural light and nature were suggested by painting plant 'shadows' on the walls and floors where the shop furnishings also cast their shadows. These and other careful touches let customers experience the store's intrinsic care and quality. Its name makes reference to Japan's international dialling code (+81).

Loran & Co Office
Loran & Co

An interior design project for the offices of a gym and spa management company. The main feature was a single, elongated desk, intended to create a multipurpose working space within an otherwise untouched room. Thereby, the company's corporate motto – 'working the body's inherent potential to the maximum' – was applied to its offices as well, awakening the possibilities inherent in the smallest unit of a workspace: the humble desk. One item thus forms almost the entire office, serving as a work surface, as well as the floor and walls; it even extends to form a conference table and the CEO's office space.

Tile-Plant
Lixil

We tend to associate the gaps between tiles with negative things like dirt and mould, but this design turns those spaces into something positive. By creating a grouting material that can cultivate plants such as grass, Tile-Plant presents an opportunity for tiles to be used in an entirely new way.

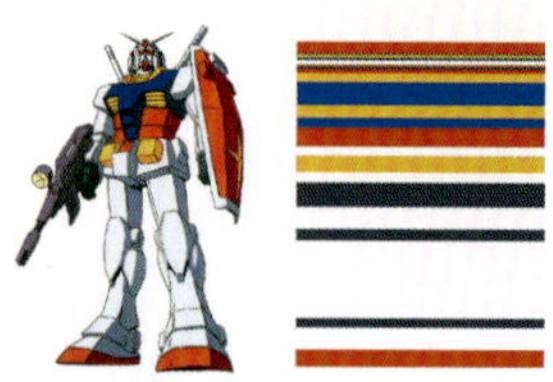

Gundam
Suntory Museum, Tempozan, Osaka

The 1980s Japanese television cartoon 'Mobile Suit Gundam' gained an immediate following among the country's young boys. This design was for a group exhibition, held at the Suntory Museum in Osaka, by Japanese artists who paid homage to its influence on popular culture. Drawing on the distinctive colours sported by the mobile suits, a set of 'go faster' stripes was developed that was intended to express the characters moving at high speed. This motif was used on everything from exhibition graphics to products sold at the venue's gift shop – and was an abstract way of referencing the cartoon without depicting the characters themselves.

Trinos
Title

A lighting fixture made from flat pieces of Japanese rice paper, woven roughly together like twigs in a bird's nest. Its construction creates a unique interaction between light and shadow. The material is laminated with plastic resin, making it hard to tear and flame resistant – properties that are perfectly suited to this seemingly fragile design. While far more durable than traditional *washi* paper, its translucence creates a quality similar to a Japanese lantern.

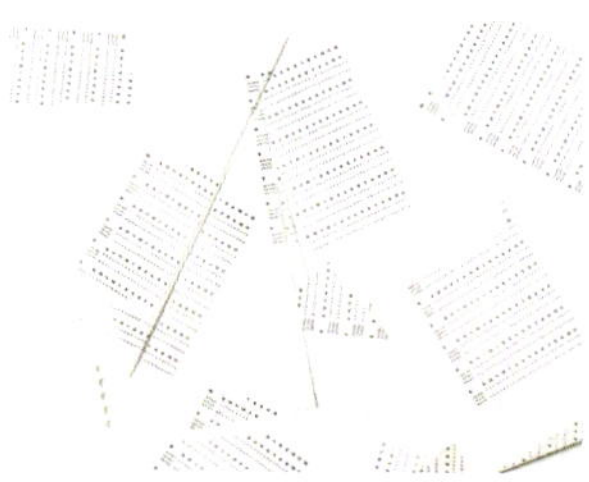

Stamp-Diary
Ryohin Keikaku

Sometimes, the appearance and size of a daily personal organizer takes precedence over its functionality. But if the schedule pages are created using stamps, it can be designed in any way the user likes. Applied on large pieces of paper, the stamps can be assembled freely to create calendars and project charts.

Luxe
Luxe

A hair salon design based on an alternative concept of luxury. The aim was to create new expressions of quality by tipping the balance and concentration of costs. Instead of splashing out on every aspect, the budget was targeted on specific elements, while the rest were treated more modestly.

Rebondir
Senshukai

A shop design for women's mail-order clothing brand Rebondir. The 'printed' space aesthetic aimed to remind shoppers of the brand's origins in the printed-catalogue mail-order market, with its suggestions for coordinating outfits. A wood-grain pattern was printed onto stone floor tiles and a stone pattern on timber furnishing items. This created an intriguing mix between the warmth and softness of wood and the solidity and massiveness of masonry.

Equbo
Tsubame Shinko Industry

The taste of good food can bring a smile to your face, which may in turn reveal your dimples. The Equbo cutlery set aims to capture this moment with its very own dimple features, embossed into the metal surface. By adding one small detail, the cutlery demonstrates that even the simplest of objects can inspire affection in a product's users.

Seesway
Seesway

A retail space design for street-fashion brand Seesway. Street fashion draws on different styles regardless of genre, and creates new value and individuality from them through reuse. The interior design suggested ways of adding functionality and value to a basic stool by duplicating and arranging it in unconventional ways. Lined up flat against the walls, stools form perimeter seating; piled upright, they turn into a sales counter or shelving elements.

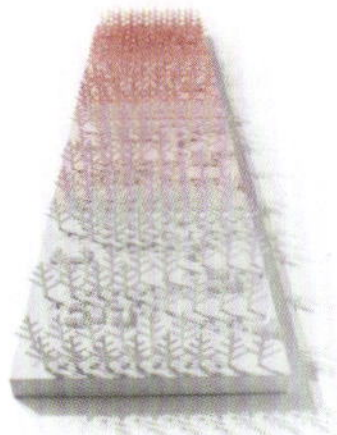

Litmus-Garden
Brutus

A miniature garden created in collaboration with the Japanese culture-and-lifestyle men's magazine *Brutus*. In reaction to contemporary environmental issues, Litmus-Garden acts as an indicator of acid rain.

Using litmus paper, which changes colour according to the PH levels it absorbs, the garden turns pink when exposed to an acidic substance, therefore alerting people to the presence of contaminated rain. This invisible environmental phenomenon is easily overlooked, but the garden renders it literally visible, helping us to rethink our relationship with nature.

Hana★Calendar
Take Art Collection

A calendar that uses the motif of falling petals to suggest the passage of time. Each week is represented by a flower with seven petals, which can be taken out of a larger bouquet and placed in a cup or a vase as you would a normal flower. Users then 'pluck' one petal for each passing day.

Fireworks House
Private Commission

A residential housing project located in the city of Chichibu, Saitama Prefecture, known for its annual December fireworks festival. It was designed with the intention of allowing the elderly client to be able to watch the display comfortably from home.

Its utilities are concentrated in the centre of the ground floor, and are surrounded with open-plan living space that is wheelchair-accessible. This allows the upper level to be focused entirely on the view. In order to optimize this capacity, the direction of the house was taken into consideration and a large, glass window installed in the roof.

Inakappe Noodles
Yokose

An interior design project for a restaurant that prides itself on the skill of its staff and the freshly boiled, handmade noodles it offers. An open kitchen lets customers see chefs at work, but here, greater intimacy is achieved by using the device of a chef's table – at which food is created especially for those invited into the kitchen, who can fully experience the energy and warmth of the cooking process.

Shoki
Tokyo Design Week

A collection of ceramic containers whose tops have been sliced off at three different heights to create a vase, a cup and an ashtray. Inspired by the informal action of reusing finished drinks cans as ashtrays, this design transforms a single familiar shape into a product with multiple uses. The name, Shoki (first instrument), alludes to the idea of the empty can as a primitive object.

Yori
Tokyo Design Week

When two people sit down on Yori's slightly twisted seat, it causes them to lean imperceptibly towards each other. A tiny variation in form thus brings people together. Research suggests that incorporating a subtle relief into an otherwise flat seating surface alters the psychology of the sitters – encouraging more social interaction.

Take Art Collection Exhibition
Spiral

All the products at this 'art department store' exhibition were limited editions, with each reasonably priced item on sale for the duration of the event. This unusual combination of art show and retail space was brought out in the exhibition design; products were displayed on stands made from rolls of bubble wrap, embedded with light sources and held together with craft paper. These elements were then disassembled into sheets and used as packing materials for the sold art pieces.

Uehi! House
Private Commission

A Tokyo residence built for a married couple of Italian-language instructors with two children. The clients required communal space for lessons, film screenings and parties, as well as private space for family relaxation. They also sought a connection between these two types of space.

On its tight plot, the house's smaller rooms were arranged around two large salon and dining-room spaces, which had 4.7-metre (15½-foot) ceilings. Twin spiral staircases allowed free movement, creating that sense of connection. Windows were cut out between rooms, so that occupants could call *Uehi!* ('hey!' in Italian) to each other – just like they were in the backstreets of Naples.

2006–7

Illoiha 1st Concept
Loran & Co

An interior design and brand development project for a fitness club near Tokyo's Ebisu Station. The intention was to express the idea of fitness and well-being achieved through physical movement. For the gym's main ceiling area, a 50-metre (164½-foot) corridor, a film was sourced that appears transparent when seen straight on, but translucent when viewed from an angle. It was layered over a long textile, so that customers could only see the true beauty of the fabric's pattern when moving down the hallway.

Acuo
Lotte

A brand-development project for a variety of breath-freshening gum. The logo features a circular shape in the centre of the 'O', which could be interpreted either as an open mouth or a water droplet. Avoiding overdone expression, the placement of the logo's letter-forms instead suggests cool, pharmaceutical expertise. Additionally, the packaging design employs gradations of metallic colours to emphasize a brisk, minty sensation and fresh breath.

Chichu-Cha
Chichu Art Museum, Naoshima

A packaging design for two leaf teas that also retail at the Chichu gift shop. The first type is relaxing, so was named Chichu Iyashi-Cha (Soothing Chichu Tea). Its canister is a soft brown, like the tea, and spreads at the base into a relaxed, comfortable form. The second tea, called Chichu Socha (Invigorating Chichu Tea), has a stimulating effect – so its canister inverts the previous shape, 'standing to attention', and shares its tea's green colour. A third white, straight-sided container holds both types of tea.

Bend
1% Products

A series of baskets created by bending alternating strips of metal. Arranging the strips one way forms the bases of the containers; the other way forms the 'walls'. The baskets arrive flat-packed, allowing the customer to decide on their preferred depth and diameter.

Bubbles
1% Products

A coffee table design in which circles of black steel cling together and break apart within a larger circular frame, resembling bubbles on the surface of a body of water. As a result, the table has a light, floating feeling that could cause a moment of hesitation when placing something on it, for fear it might fall through the transparent 'watery' surface.

N702iS
NEC

A mobile-phone handset made in collaboration with NEC. For a feeling of accessibility, the design drew on the physical attributes of a drinking glass – an equally familiar, hand-held form. The phone was cast in two layers of transparent and coloured resins, and a variety of inner and outer casings made, allowing users to personalize their handset. Continuing the drinking-glass theme, a triaxle acceleration sensor was used to achieve the handset's liquid-like interface; indicator lights signalling incoming calls or texts animate it like bubbles in a fizzy drink. Additionally, data can be exchanged between handsets using a simple tilting action that resembles the act of pouring liquid from one cup into another.

Idea Frames
Idea International

An interior design project for a retail space in the Omotesando Hills shopping district of Tokyo. The idea was to create a series of frames in which to present the wide variety of products and brands on sale. In order to achieve this, thin pieces were sliced from the ends of plywood sheets of different thicknesses, and applied to the edges of shop fixtures to create more defined borders around the items on display. The varying thicknesses of these end pieces added a feeling of depth and texture to the interior.

Mado (M-Out)
Gallery Tagboat, Tokyo

Hanging a picture can be like cutting a new window into the wall of your room; the Mado frame adds to this concept by offering a new kind of view. Picture-frame corners usually connect at 45 degrees, but here they meet perpendicularly, like in a window frame. Additionally, there are three filters to slip over the picture – 'cloudy', 'window pane' and 'rain' – which are intended to reflect the experience of looking out of a real window at its ever-changing views.

Sasae
Gebrüder Thonet Vienna

A coat rack that only becomes functional when propped up against a wall. As a result, it does not over-assert itself in the room. Instead, this apparent weakness allows it to melt into the background, as though it had always been there.

Kisala
Kisala

Tokyo's Kisala restaurant comprises four areas: one counter each for Japanese and Western food, a cigar bar and a private dining room. Each space was given a different feel by varying the tone of the wood used on the floor, ceiling and walls, which shifts from dark to light. The colour of the wood panelling blends gradually between each area to signal a gentle transition. Mirrors, cut to the same size as the wood panels, are interspersed throughout, creating a feeling of interconnectedness between the distinct rooms.

Champagne-Vase
Veuve Clicquot

A vase designed for the Veuve Clicquot charity exhibition, organized by the eponymous champagne maker. Its design was derived from the fabled clarity and fine bubbles of company founder Madame Clicquot's champagne – both regarded as revolutionary for their time.

Thousands of fine air bubbles generated by laser formed a 3D impression of a vase within a block of acrylic – the exact model being reflective of a vase found during Madame Cliquot's era.

Chu-Keshi
Chichu Art Museum, Naoshima

An eraser designed for the gift shop at the Chichu Art Museum at Benesse Art Site Naoshima. It can be sad to see the corners of a new, snow-white eraser wear down and become dirty, so this version already feels lovingly used, like a secondhand T-shirt or a pair of vintage jeans. With its pre-existing sloping angles, the Chu-Keshi eraser is less likely to start feeling old or worn out.

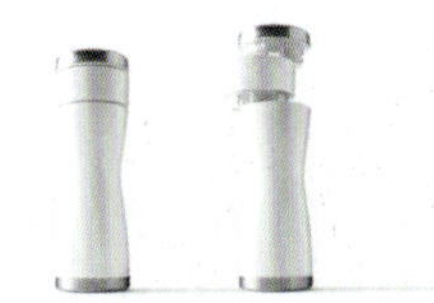

Teapot Bottle
Shizuoka Tea Commerce and Industry Association

A thermos that gives more people the chance to enjoy home-brewed green tea. The design features two compartments, one at the top for tea leaves and one underneath for hot water. Before drinking, the user can flip over the flask to infuse the leaves according to their taste. Once upright, the section holding the leaves can be removed to access the tea below. These features function as a built-in timer, so the thermos was given an hourglass shape to recall traditional English sand egg-timers.

ABC Kitchen Studio
ABC Cooking Studio

The ABC Kitchen Studio provides highly specialized culinary

services – including a range of cooking classes, recipe and meal delivery, and facilitites for catering preparation – all in one space.

The interior design reflected the kitchen's lively atmosphere through its random arrangement of sixteen different wallpapers across the walls and ceiling, all varying in pattern. The selection of wallpaper panels is reminiscent of a sample swatch book, and serves as a visual metaphor for the variety of activities and services provided by the space.

Keiji-Vine
Benesse Art Site Naoshima

A message board placed in the window of the Honmura Lounge and Archive information space at Benesse Art Site Naoshima. It needed to be accessible both inside and outside – and cover as little glass as possible while accommodating everything from large maps to postcards.

Velcro material was cut into a vine shape and fixed to the window, then Velcro hook fasteners, fashioned in the shape of flowers, birds and insects, held the various messages in place. It is broadly mirrored on each side, so that the view into the space is not blocked. The design's name combines the Japanese *keiji* (announcement) with the English 'vine'.

2008

Radial
Council

A bookshelf whose radial array of uprights makes it seem distorted, as though viewed through a wide-angle lens. Depending on their placement, books can show off their covers as well as their spines, which opens up new possibilities for display. Placing alternating, reversed sets of shelving next to each other in a row creates a horizontal wave pattern. The bookshelf's arrangement strengthens it structurally, making a backboard unnecessary and allowing its grid to frame different views depending on the viewer's position.

Green Tumbler
Kokuyo

A series of indoor planters designed for the office. Each piece combines the rounded shape of handmade pots with the straight lines often found in office furniture and equipment. The planters are all hydroponic for cleanliness and maintenance, and come in three sizes – short (desktop), tall (cabinet) and 'grande' (floor-standing) – each with a stand. Users water the short planter by opening its lid, as though brewing coffee (it even has a 'coffee-jug' handle for portability).

Tuta
Cappellini

A coat rack inspired by *tuta* (Ivy), which grows by clinging onto trees and buildings. With its hollow, laser-cut metal form, the design dispenses with any such solid support, leaving only an outline that seems to float in space.

Kenzo Parfums
Kenzo

A retail-space design for Kenzo Parfums, which was launched in Paris and then rolled out in 250 stores across 80 countries. As a French brand with roots in Japan, Kenzo's image is one of dualities – it is this characteristic that inspired the concept for the design.

To generate the main visual motif, Kenzo's logo was deconstructed and its individual pieces rebuilt into random decorative arrangements, similar to twigs in a nest. By using contrasting materials – lacquered white steel was paired with dark wood – the brand's multifaceted identity was further emphasized. Coordinated counters, stools, lamps and mirrors continued the theme of the space.

Flower-Scale
Guzzini

A contribution to *Foodesign* – an exhibition organized by Italian home-goods manufacturer Guzzini. The aim was to use design as a means of re-engaging the Japanese custom of socializing over food, which is becoming less common as more people eat alone.

'Bouquets' of flower-shaped cutting templates for slicing edible treats invoke the concept of sharing food by helping users to divide them (odd numbers like five, seven or nine, which are hard to cut unaided, deliberately reinforce the need to use Flower-Scale). Diners hold the shorter, flower side when cutting deep items like cake, and the long stem when slicing thin pies and pizzas.

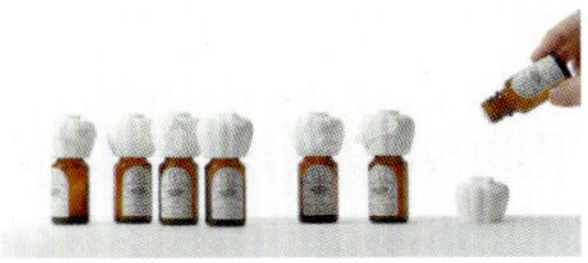

Aromamora
Aromamora

A bottle design for a line of essential oils blended especially for each season. Essential oils often require a diffuser for their fragrance to be enjoyed. Aromamora, however, combines the two products into one design. Its cap was fabricated through powder sintering rapid-prototyping (RP) technology for a porous structure that releases fragrance efficiently and effectively. With a shape recalling a magic pumpkin and features like a zip, padlock or secret drawer, the cap/diffuser allows Aromamora's 'world' to unfold, one scented moment at a time. The design of the cap changes according to the seasonal scent of each oil.

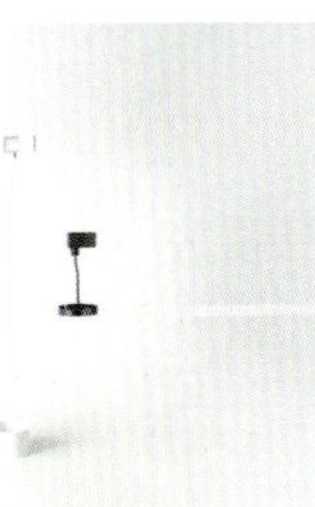

Phone-Stand
Au

In the evening, after work, a mobile phone isn't mobile. At home, you need to be able to answer it easily, recharge it or take photographs (for which stability is preferable). Phone-Stand integrates all these capabilities with an outlet for any handset specification.

The base of the taller, floor-standing version doubles as a cord holder, and that of the tabletop variant is deep enough to hold keys and small change. Additional digital content can also be applied, such as a timepiece and an animated on-screen flame, transforming Phone-Stand into a 'candelabra' or 'clock'. Rather than setting your mobile aside, this design can make it the focal point of the room.

Asobio 1st Concept
Cobest

An interior design for Italian fashion brand Asobio's flagship Shanghai store. Drawing on the concept of a photo studio, and the idea of photography as a reflection of reality, the garments were presented as photo-shoot subjects; framed by 'lightboxes', they were hung from racks with camera-tripod legs and lit by studio-style spotlights. Two types of mirrors lined the walls: 'true' ones, showing viewers' reflections, and 'false' ones through which people could actually pass into other, seemingly hidden, mirrored spaces.

2009

Blown-Fabric

Tokyo Fiber '09 Sensewear

Designed for the exhibition *Tokyo Fiber '09 Senseware*, held during the 2009 Salone del Mobile, Milan, Blown-Fabric was intended to convey the possibilities of new materials developed with Japanese synthetic-fibre technology.

'Smash' is a specialized, long-fibre, non-woven polyester that can be manipulated into different forms through heat-press forming technology. The material is thermoplastic, light and tear-proof, and glows beautifully when light passes through it; qualities that made it ideal for the creation of lighting fixtures in the style of vernacular Japanese *chochin* paper lanterns.

The structure of standard *chochin* comprises thin strips of bamboo wrapped around a wooden frame and strengthened with vertical stitching. Japanese mulberry paper pasted over the frame completes the lamps, giving them their characteristic glow. However, Smash's particular properties allowed it to be shaped like blown glass into a seamless, one-piece lantern. It is impossible to completely control this process, so each fixture takes a unique form as heat is added and pressurized air blown in. As in glass-blowing, makers can intervene during production, resulting in objects with infinitely varied imperfections – a far cry from the standardized forms of industrial mass production.

124

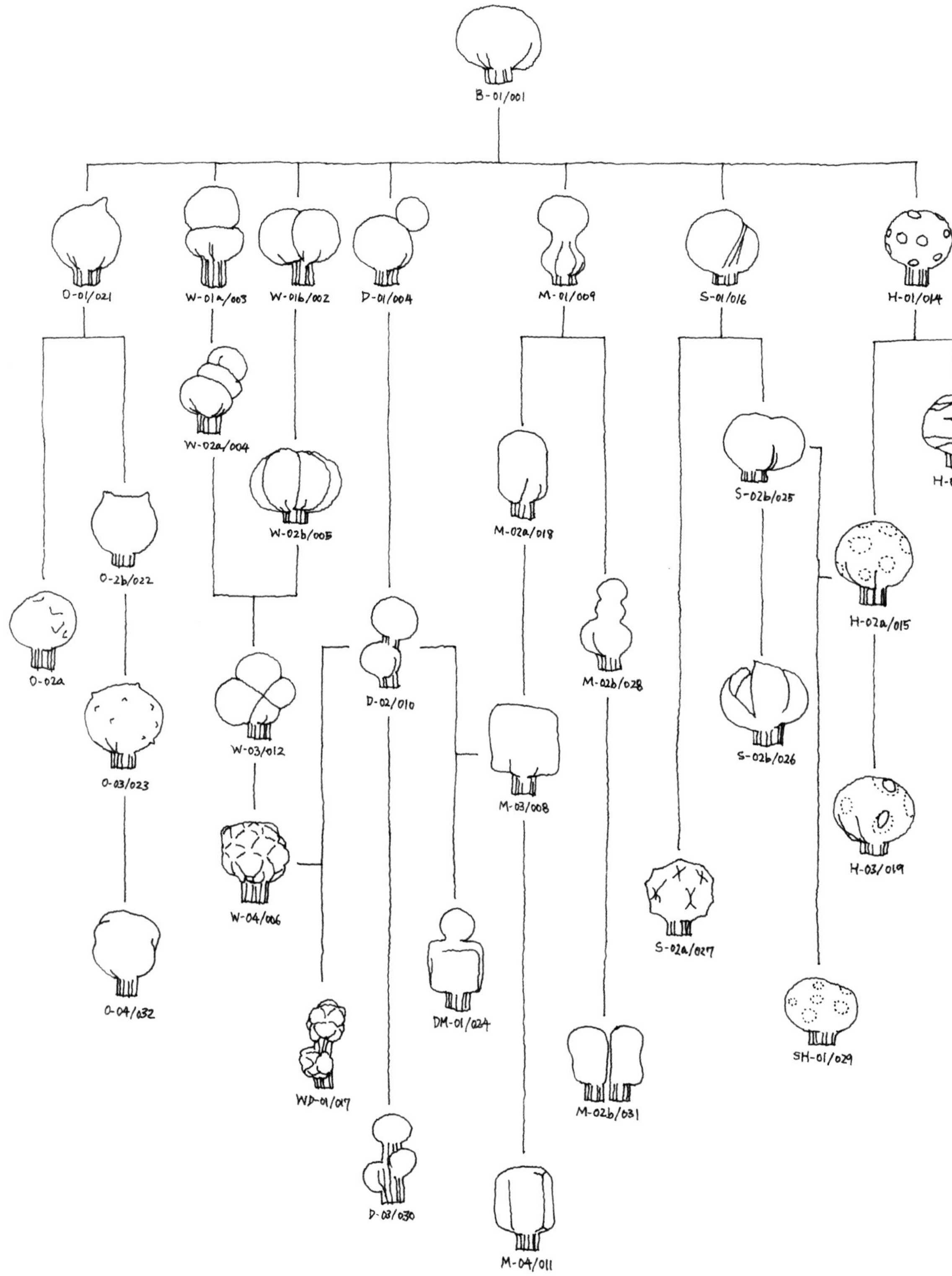
B-01/001
O-01/021
W-01a/003
W-01b/002
D-01/004
M-01/009
S-01/016
H-01/014
W-02a/004
W-02b/005
O-2b/022
O-02a
O-03/023
O-04/032
W-03/012
W-04/006
WD-01/017
D-02/010
D-03/030
M-02a/018
M-02b/028
M-03/008
DM-01/024
M-02b/031
M-04/011
S-02b/025
S-02b/026
S-02a/027
H-02b/020
H-02a/015
H-03/019
SH-01/029

Target

Arketipo

A bookshelf that replaces the usual horizontal shelves with an ordered array of small crosses. These work both as shelves and as bookends, allowing even thin magazines or a small number of volumes to stand on their own. Since the shelves are not unidirectional, the units can be used any way up.

Moya
Arketipo

If a surface doesn't cast a shadow, does it exist?

The Moya (haze) low tables combine a glass tabletop with a base that has been hollowed out to form an exact half-sphere. The result is that the top seems to lack shadows and almost merges with the base – with the shape beneath it defying any sense of depth or perspective, much like a haze in the atmosphere. This illusion only dissipates when an object is set on top, casting a shadow and bringing the depth of the table into focus.

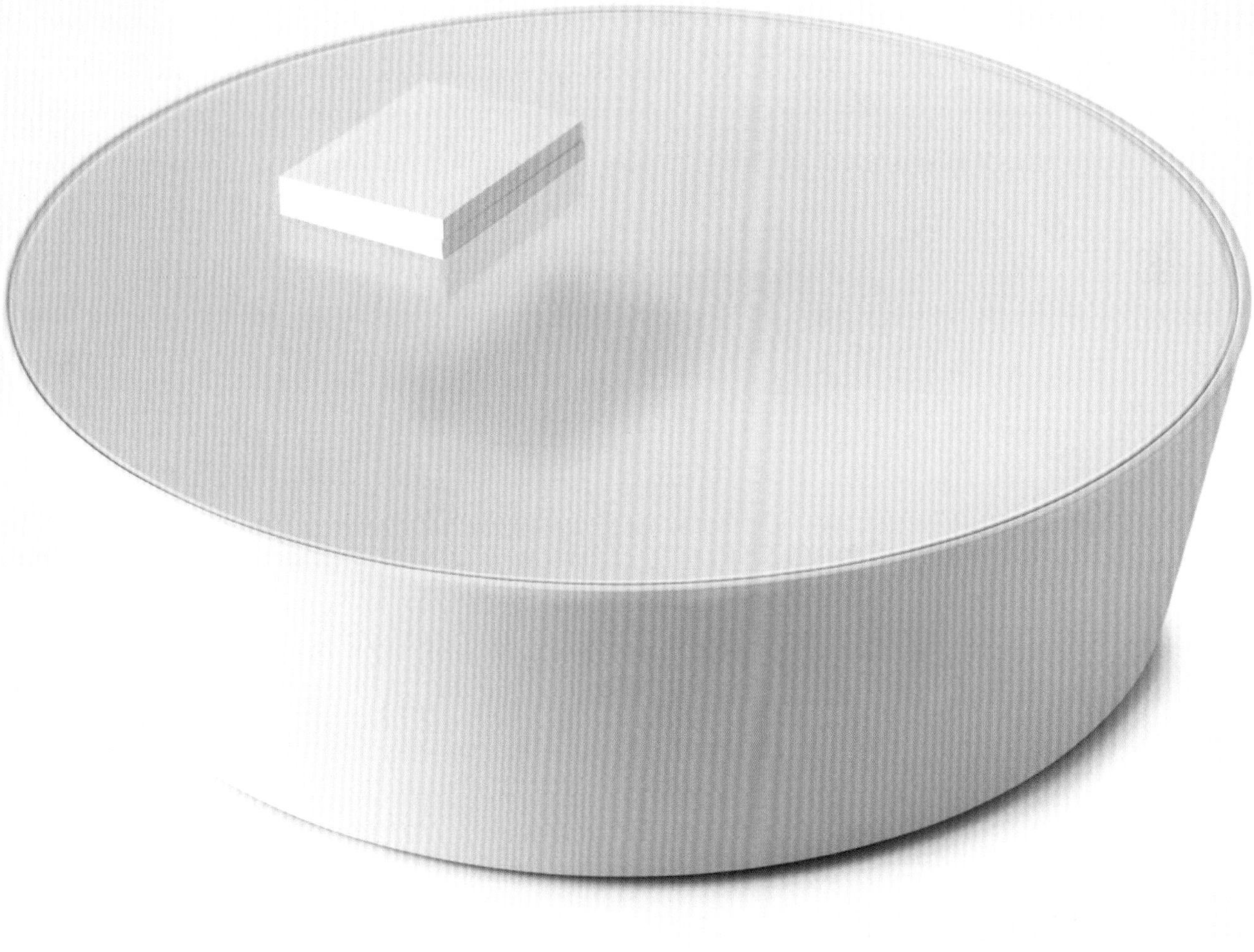

24 Issey Miyake 1st Concept

Issey Miyake

A new, high-concept shop offering a brightly coloured selection of items chosen from each of the Issey Miyake lines, alongside new, original products specially created for the store. Each reasonably priced item comes in twenty different colours, and the shop's lineup is renewed every two months. The idea behind the design derived from that of the *konbini*, the Japanese 24-hour convenience store, with its constant state of dynamic, fluid change – what could be termed 'harmonious chaos'.

To highlight this association, the shop's name is '24', and its logo features the kind of graphic stripes that a customer might expect to find on the signage of such an outlet. The plastic packaging, too, takes its cue from the convenience-store aesthetic. To keep the space as compact as possible and fill it to the brim with products, the customary back-room storage facility was removed – all of the merchandise is on the shelves at all times.

Since the products are so brightly coloured, pure white was used for the shop itself. All of the fixtures, including hangers and shelving, are constructed from 7-millimetre (1/4-inch) -diameter steel rods, similar to a wire shopping basket. The entire shop functions as a display, and the white lines of the steel fixtures accentuate the vivid hues of the products on show. The idea is that as new stock arrives to replace the old, the shop itself will change character dramatically.

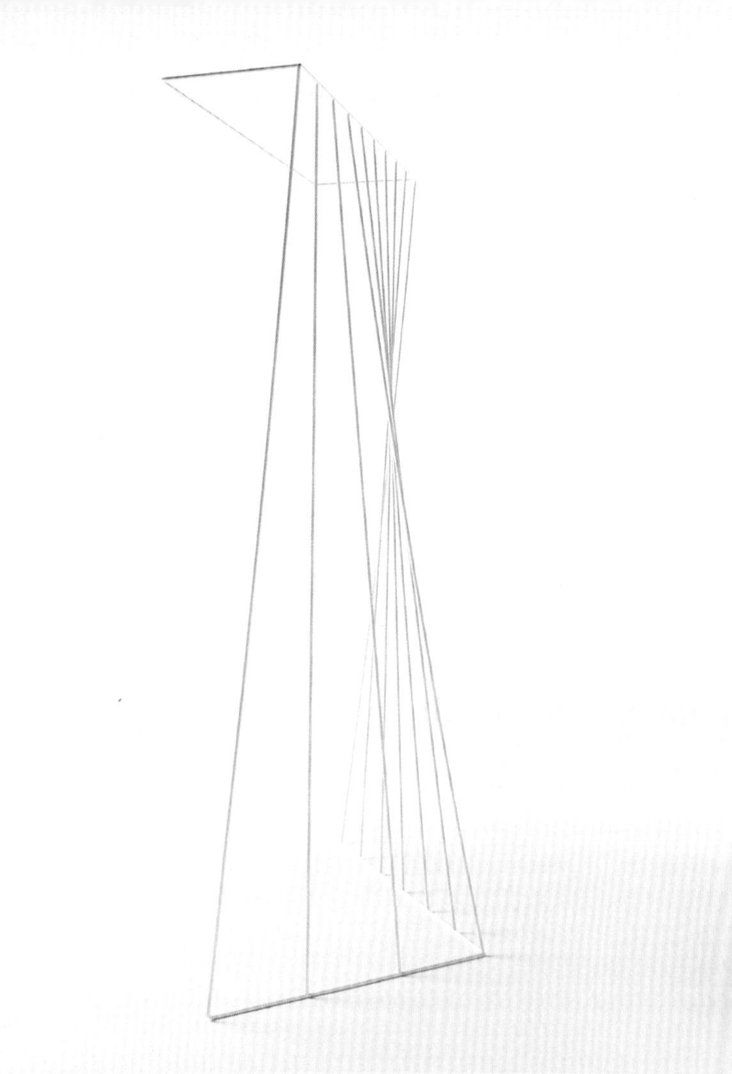

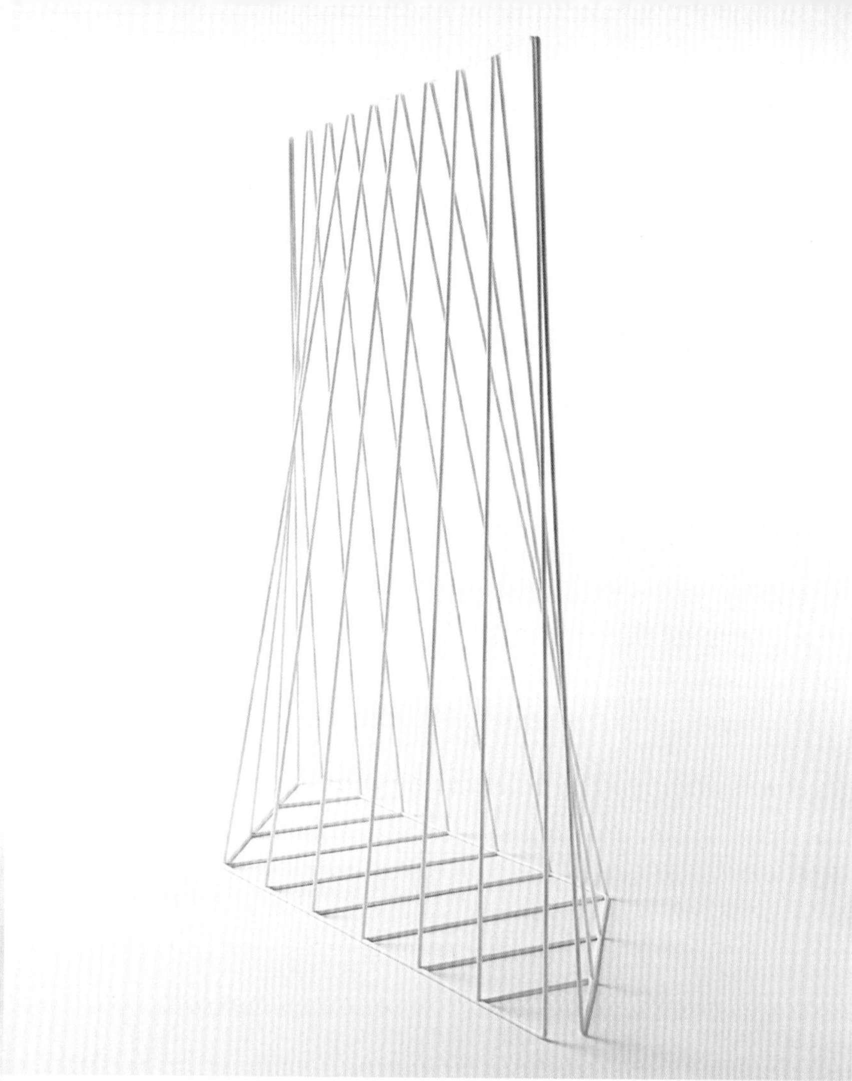

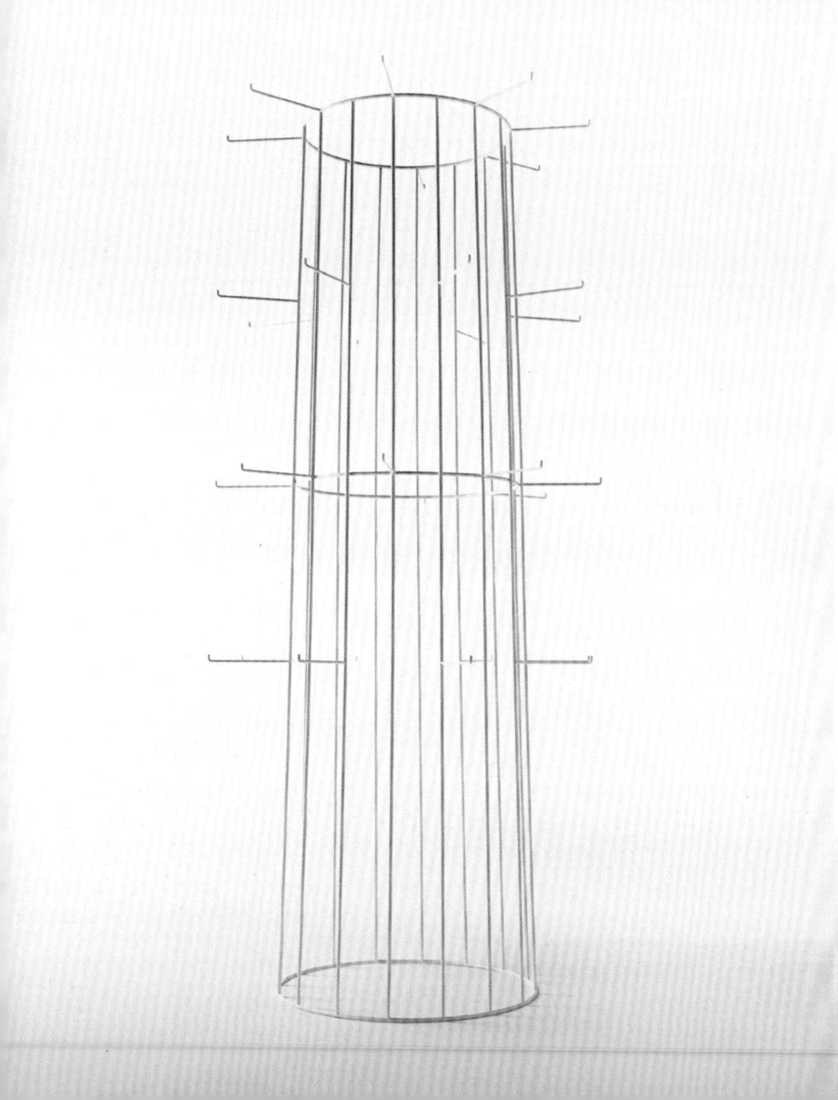

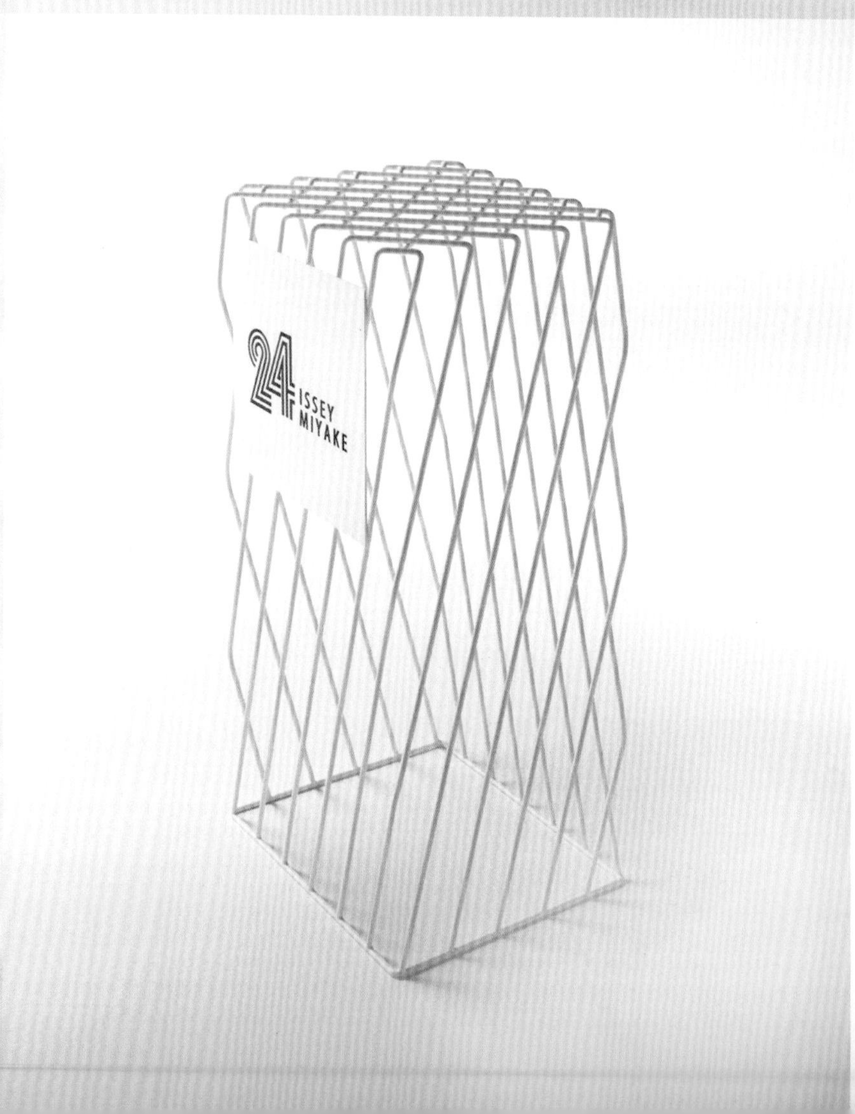
24
ISSEY
MIYAKE

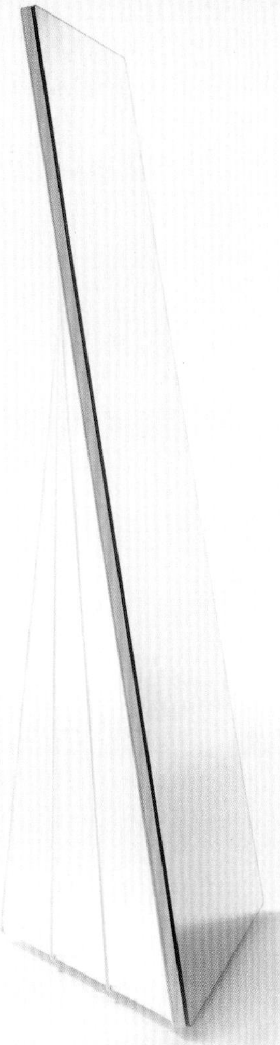

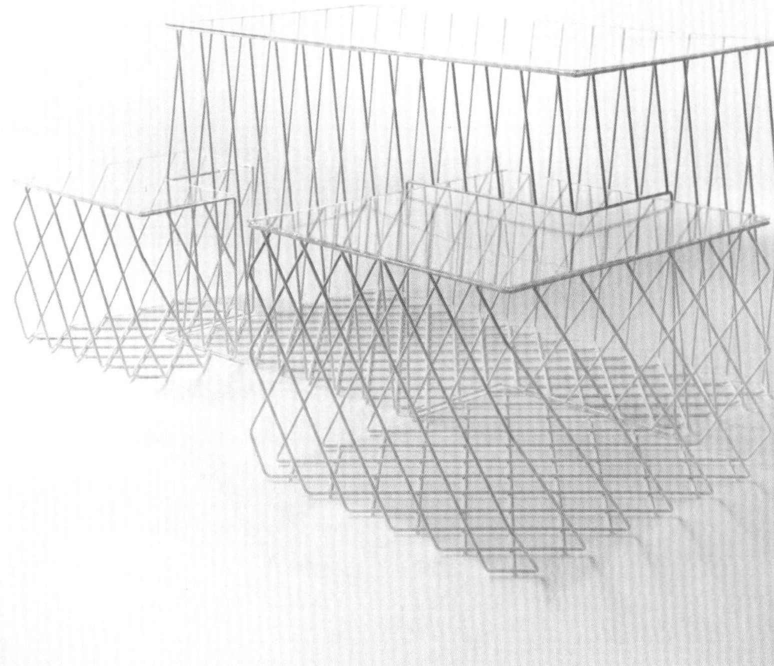

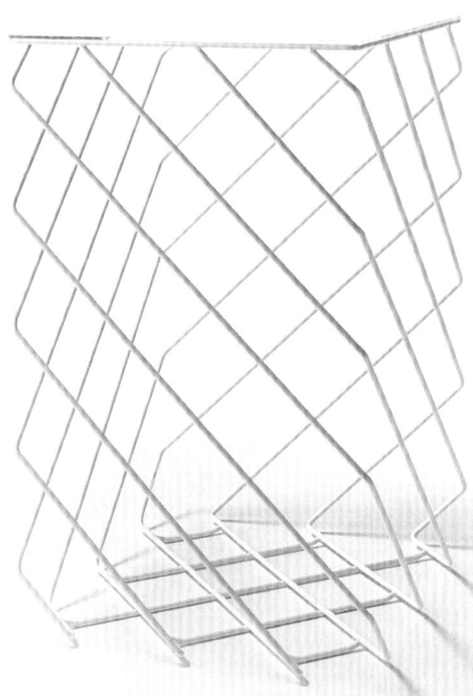

Notescape

Moleskine

A contribution to Moleskine's *Detour* exhibition, in which the venerable Italian firm gave fifty designers and artists free range to create a piece for the show.

For a designer, sketching is often the means by which early ideas are formed, or final ones finessed. Sketches can be used to consider a product's overall form or to refine last-minute details. Always present in the development of a project, sketching is free from the constraints of physical construction, and can lead a person on an unpredictable journey with unexpected (and infinite) outcomes. Representative of this adventure, Notescape was designed as a vast, never-ending landscape. By cutting the pages of the sketchbook to create this undulating, three-dimensional scene, the design demonstrates how the sketch relates to (and transforms) both two- and three-dimensional space, and presents the book itself as a physical space.

Music-Cage

Au

A wireless Bluetooth speaker that allows the user to relax with music, as though listening to birdsong from a birdcage. The design adopted a familiar shape and is equally suited to being hung from the ceiling or placed on a shelf or desktop. It can be attached to a ceiling-mounted electrical socket otherwise used for light fittings. Alternatively, it can be plugged into a wall socket for power and attached to an accompanying floor or table stand.

Dress-Up

Ceramic Japan

Period. Comma, "Quote"

Takata Lemnos

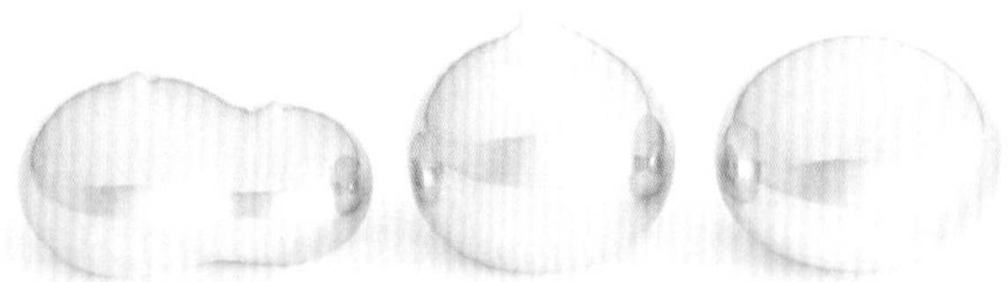

A collection of three vases that from one side appears ordinary, but when viewed from the other side reveals a different collar on each that gives the vase its identity. The three forms make up an archetypal family: a heavyset 'father', a slim 'mother' and a smaller 'child'. The position of the collar corresponds to where the flowers are placed in the vase, creating a relationship akin to that which exists between our hair and the clothes we wear. The vases are made from unglazed porcelain, further animating the objects with a clothing-like texture.

A trio of aluminium paperweights shaped like punctuation marks: 'Period' for work completed, 'Comma' for work in progress and 'Quotation Mark' for messages. Each appears as if it adopted its form naturally when the aluminium hardened during casting. Unlike with chrome plating, the soft lustre of the weights' hand-buffed surfaces displays the minuscule scratches and tarnishing that occur with use – but it is easy to restore their sheen through polishing. The paperweights were designed with this activity in mind, to give owners a way of exploring and enjoying the particular charm of aluminium over time.

Corona

Watanabe Kyogu

A collaboration with the company Watanabe Kyogu, which makes high-specification, hand-crafted globes in Soka, Saitama Prefecture, just north of Tokyo. By reducing the amount of information normally found on a globe, and rendering the oceans in plain white and the land masses in black, the Corona globe highlights the outlines of islands and coastlines. In this way, it aims to be a conceptual, rather than simply informative, learning aid. Often visual design features that are depicted in high contrast are discouraged, as imperfections in the finish can be more conspicuous. However, with Corona, these details were possible (and perfectly executed) due to the cutting-edge technology and highly skilled craftspeople involved.

Peel

Ceramic Japan

A small section of the rim of this tea mug has been 'peeled' away. This allows a teabag string to escape, and/or a spoon to remain inside when the user drinks their tea or puts on the lid in order to carry it. The lid helps to steep the tea; it can also be used as a small dish for sweets or biscuits, or to rest a used teabag or a spoon on. The bases of the cup and lid are indented for easy stacking and storage.

MD.net Clinic

MD.net Clinic

An interior design for a mental-health clinic in Akasaka, Tokyo. Rather than getting patients back to a 'zero', a neutral starting place (the traditional model in mental-health care), MD.net Clinic aims to provide them with something extra: a further richness in their daily lives that they did not have before starting treatment.

The design represented an attempt to express this philosophy in spatial terms. The 'doors' that line the walls of the clinic do not open, instead, seemingly ordinary parts of the walls open up into new spaces. The consultation rooms are entered by sliding the bookshelves sideways. The door at the end of the hallway opens on to a window; the amount of light in the hallway is controlled by opening and closing this door.

By providing alternative perspectives for viewing the world, and avoiding being trapped by pre-existing perceptions, the interior allows visitors – and staff members – to open their minds and discover new ways in which to approach the challenges of their lives.

Phantom-Waves
Museum of Arts and Design, New York

Polarizing films block light when they are layered at 90 degrees to each other, but allow it to pass through when layered in parallel. Phantom-Waves uses a combination of these parallel and perpendicular layers within the surface of an acrylic vase to create the visual illusion of three-dimensional boundaries. When cut into a curve the wave-shaped film appears like pools of rippling water; branches and flowers placed inside the vases appear to pass through its surface.

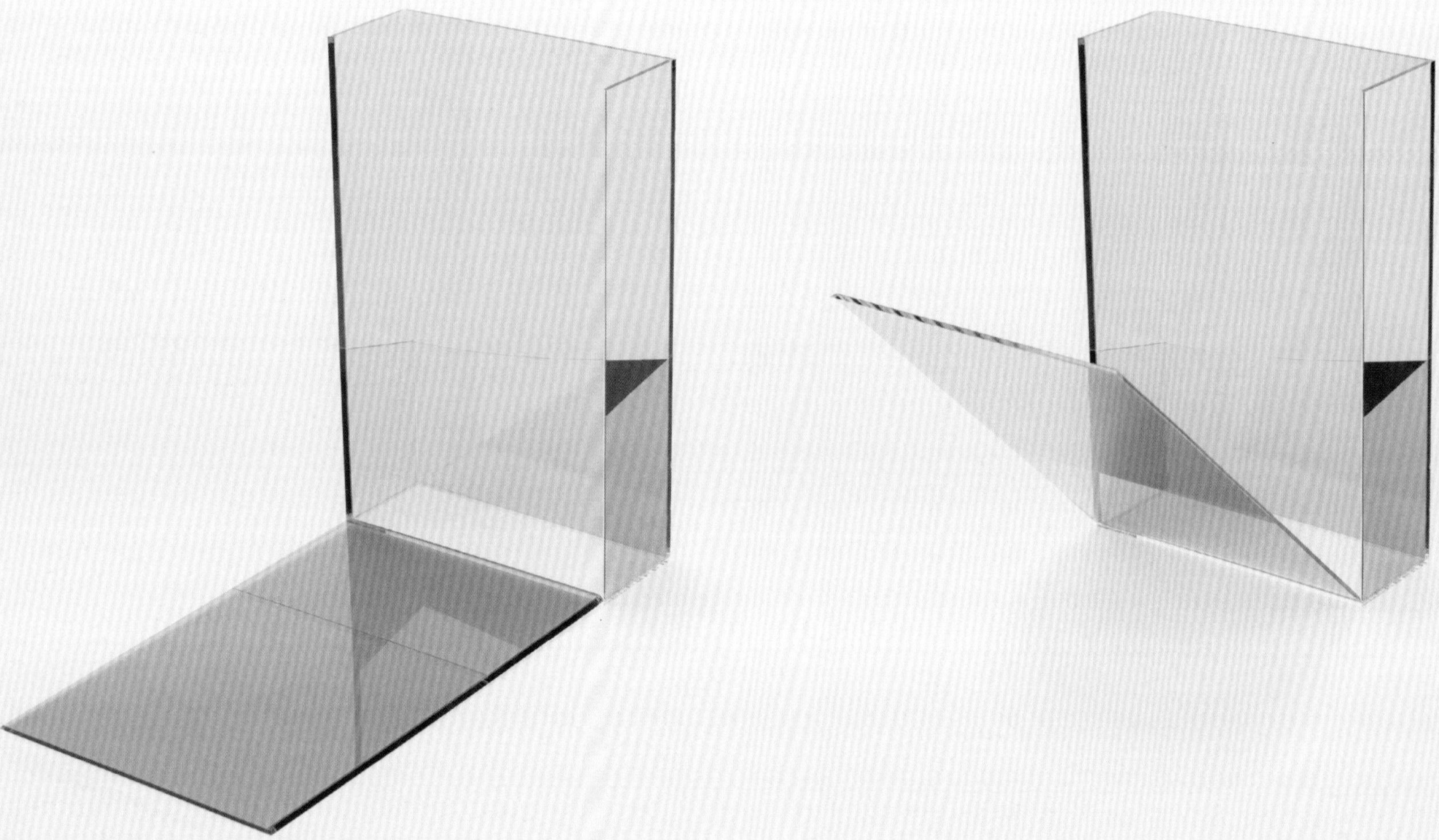

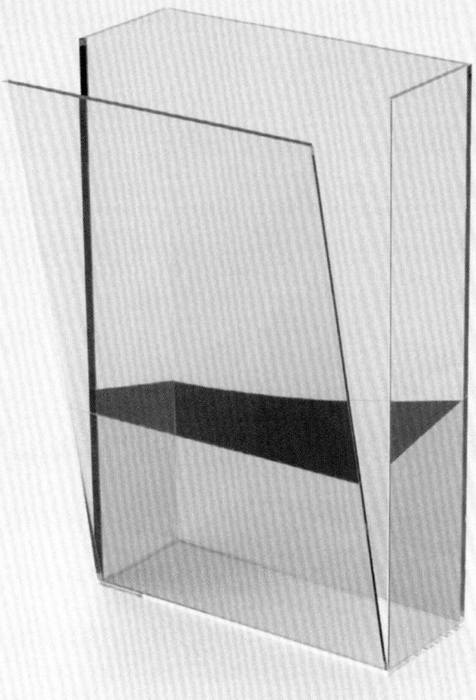

Shoe-Horn

Maruni Wood Industry

As the name suggests, shoehorns were first made by carving animal horns; this tall, conical implement pays homage to its origins. When the top part is lifted, the shoehorn comes away in a separate section. It is connected to its stand by an embedded neodymium magnet, and both parts are carved from a single block of wood for complete continuity of the grain pattern. The shoehorn can be removed with one hand as the stand is weighted for the appropriate amount of resistance.

145

International Kogei Triennale Pre-Event

21st Century Museum of Contemporary Art, Kanazawa

An exhibition design for sixty-two craft objects by fifty artisans at the 21st Century Museum of Contemporary Art, Kanazawa, Japan. The objects' materials ranged widely, with glass, wood, ceramics, metal and cloth among those featured. Additionally, a variety of techniques had been used in their creation, and the items differed greatly in size. For the installation, therefore, the opposite strategy was chosen. Small, identical, mass-produced glass cases, which are sold for home use as greenhouses, gave a sense of order to the space and provided visitors with a neutral basis from which to view the exhibition. They also allowed the rich variety of the objects to stand out, and were subtly suggestive of the museum's hopes that crafts will continue to grow and flourish like plants.

These glass display cases each came with their own shelving and pre-installed wiring for lighting. They were inexpensive, ideal for the tight budget, and easily assembled with just a screwdriver. It took only one day to put together all sixty-two of them. They could also be easily dismantled and stored after the show. This not only eliminated nearly all waste from the exhibition fixtures but also allowed for possible reuse.

Cord-Chair

Maruni Wood Industry

A chair, made in collaboration with manufacturer Maruni Wood Industry, based in Hiroshima Prefecture, with legs that are only 15 millimetres (1/2 inch) in diameter. To achieve such extreme thinness, each wooden leg was hollowed out and inserted with a 9-millimetre (3/8-inch) steel frame. Of particular interest in the design process was the opportunity to show off the flawlessness of the timber used, so special care was taken to align the grain of each respective part. Additionally, the chair's non-structural wooden elements were left simple and unadorned in order to bring out the impeccable nature of the material.

The Cord-Chair is not about the kind of 'mechanical beauty' that seeks the lowest common denominator in form and structure. Rather, with this design the intention was to explore the relationship between paired materials – like those found in electrical wiring, with its metal cord and the rubber that encases it, or the combination of reconstituted masonry and steel bars in reinforced concrete. This liberates the wood from its usual structural role in furniture – really showcasing the material and bringing out, to continue the metaphor, its 'greatest' common denominator: natural warmth and softness.

Ghost Stories 2nd Chapter

Museum of Arts and Design, New York

A solo exhibition of nendo's work at the Museum of Arts and Design in New York. Long, rectangular slashes of windows run horizontally around the outer walls and ceiling of the allocated gallery space, so a decision was made against installing any interior walls that would keep visitors from enjoying the changing light that filters through the windows into the gallery.

The exhibited pieces were presented on white plinths or directly on the floor, and were then surrounded by small, 5-millimetre (0.2-inch) -diameter white stickers, each one affixed by hand. The stickers provided a gradation effect, which visually blended the displays into the larger space as if they were slowly disintegrating. The aim was to maintain the space's specificity while intensifying the presence of each new work.

Link

Arketipo

A side table supported by two orthogonal arches joined at a single point. The form that results is reminiscent of the links in a chain, giving the table its name. It is also from this chain-like construction that the table draws its strength, which is compressive rather than tensile. Although there is only a single contact point between the tabletop and its base, the overall structure enables stability across the entire surface.

158

Eda

Arketipo

A freestanding coat stand that positions vertical metal *eda* (branches) in a large, ceramic vase filled with sand. Changing the coat stand's look is as simple as rearranging the branches. To accommodate more coats, the user only needs to add more branches to the arrangement. Silicone rings fitted inside the base act as holders for each attachment so that they do not scratch against the side of the vase.

Fadeout-Chair

Museum of Arts and Design, New York

The Fadeout-Chair's legs seem to gently disappear, as though it stands in a pool of mist or fog. The chair's back and seat are wooden, and its clear-acrylic legs are expertly painted so that their 'wood grain' appears to gradually fade away. Usually, our perception of furniture is strongly affected by the space around it. The Fadeout-Chair turns this relationship on its head, allowing us to change the look and feel of a space simply by placing a piece of furniture in it. It represents an attempt to design space itself through the nature of the objects within it.

162

2010

Chair Garden

Milan Design Week

In the imaginary world of Chair Garden, a stool grows a backrest and becomes a chair; an armrest sprouts from it, forming an armchair. In another iteration the chair expands widthways and turns into a bench, or lengthways to become a lounge chair – or even a bed. If furniture could truly grow and mature over time – like plants in a garden – then, through observing this change, we might find new ways in which to design form more naturally.

The exhibition *Chair Garden*, at Milan's Galleria Antonia Jannone in 2010, explored ways of determining form based not on function or a modular system derived from the proportions of the human body, but on the concept of 'furniture that grows' – as if in a garden.

168

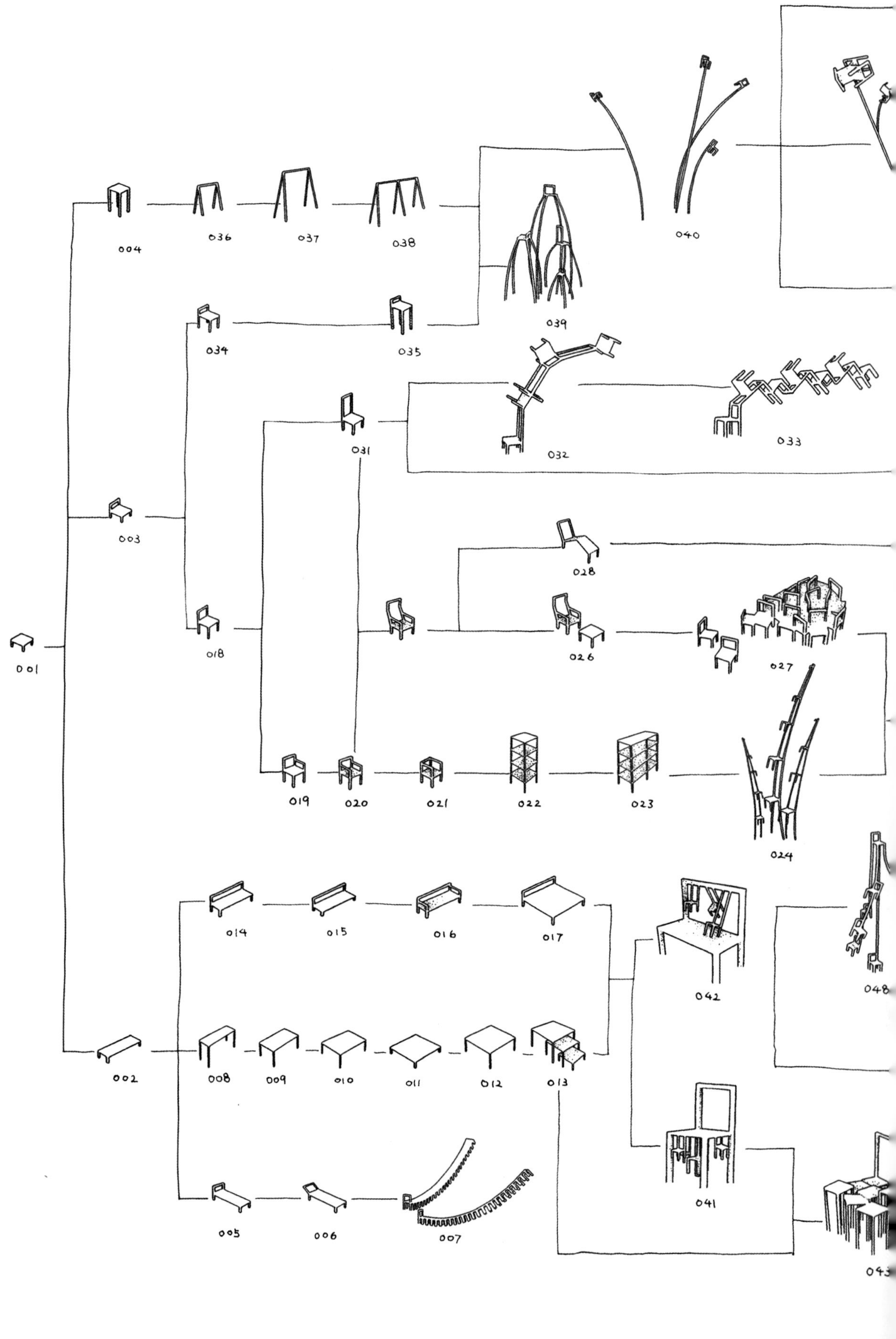
001
002
003
004
005
006
007
008
009
010
011
012
013
014
015
016
017
018
019
020
021
022
023
024
026
027
028
031
032
033
034
035
036
037
038
039
040
041
042

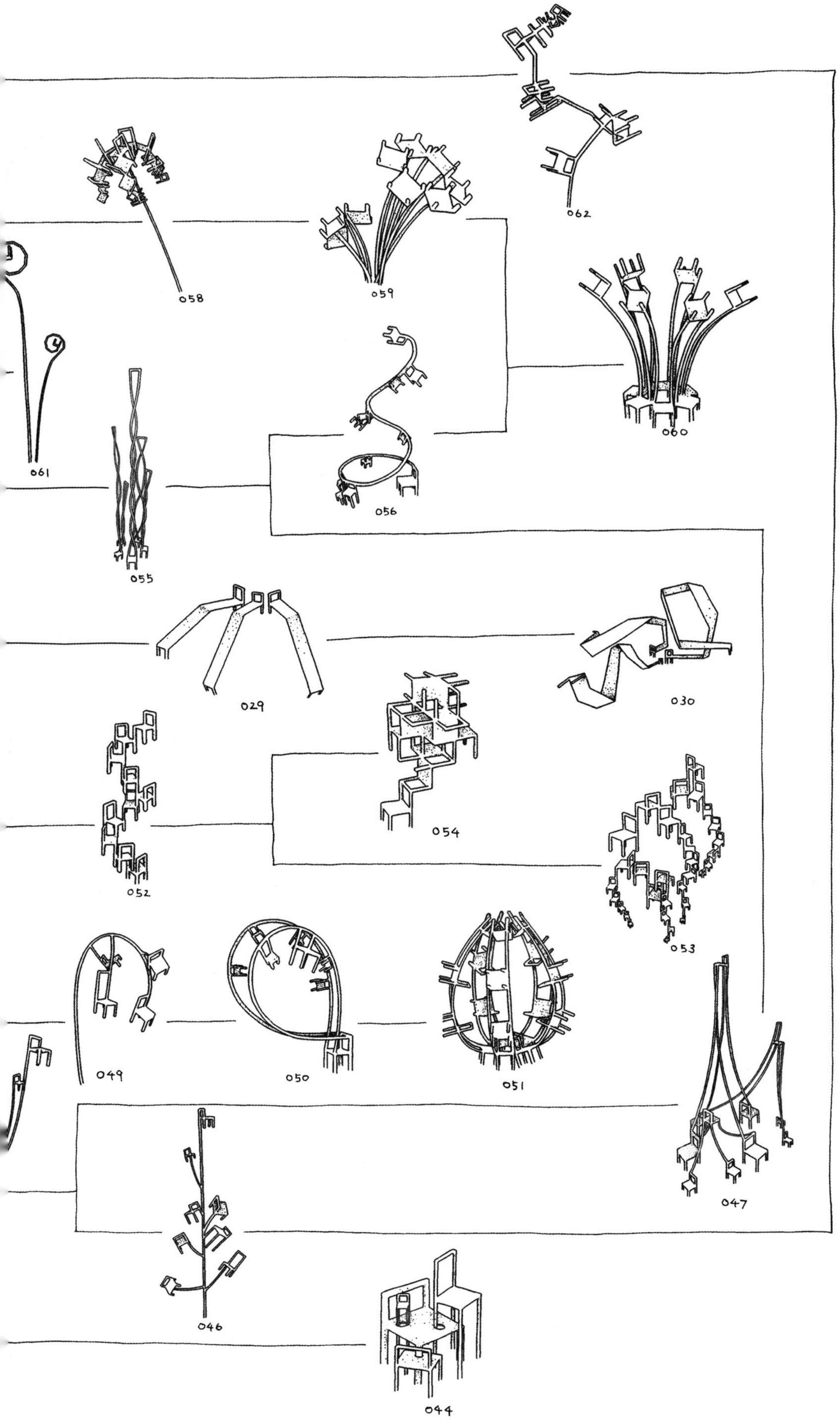
062
058
059
061
060
056
055
029
030
054
052
053
049
050
051
047
046
044

Roll

Ceramica Flaminia

Like a single piece of paper coiled into a complete circle, the Roll ceramic washbasin is less an object to hold water than a boundary around it. The ends of the 'paper' overlap, stressing its roll form but also creating a small, raised section between the two surfaces that neatly retains water in the bowl.

Othello
Boffi

A three-part storage unit with only two doors. On one side the doors are lacquered in white, while on the other they are panelled with mirrors. Their hinges pivot 180 degrees, allowing doors to be opened on one compartment and closed on the adjacent one, and with one fewer door than compartments, the user can continually change Othello's appearance. It can have an all-white or all-mirrored 'face'; additionally, its open section can be central, with one of each door type arranged to either side.

Thin Black Lines
Saatchi Gallery, London

Wiry black 'lines' of bent, solid, tubular steel, like three-dimensional sketches drawn in the air, trace out otherwise empty surfaces and volumes to which practical functions are assigned – such as hanger racks, bowls, tables and chairs. These outlines represent what remains after paring down the forms of everyday domestic objects to their basic essence. The abstract forms also recall characters in Japanese calligraphy, with their condensed expressions of meaning.

At times the items even bridge the traditional space between two and three dimensions. Multifaceted and constantly morphing according to the viewer's changing standpoint, they alternate between a 'becoming' and a breaking down of form.

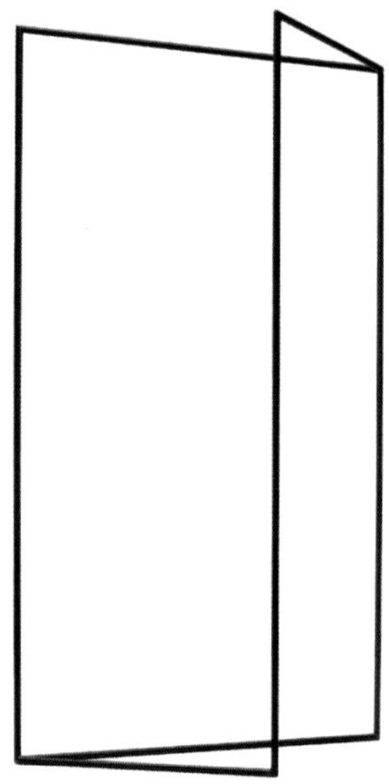

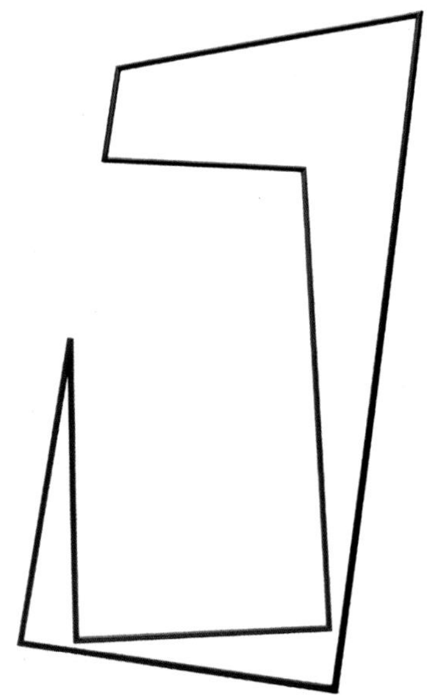

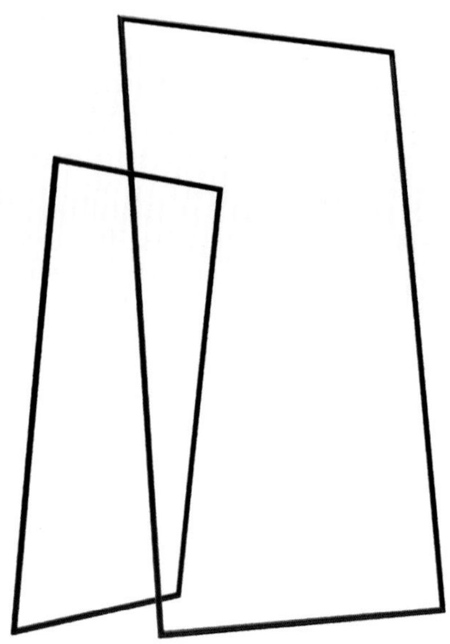

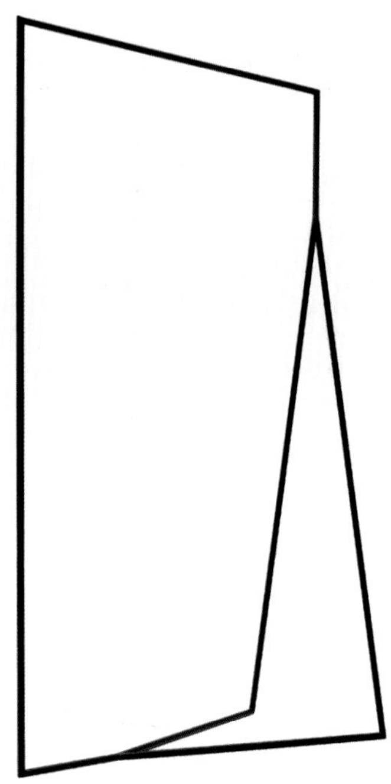

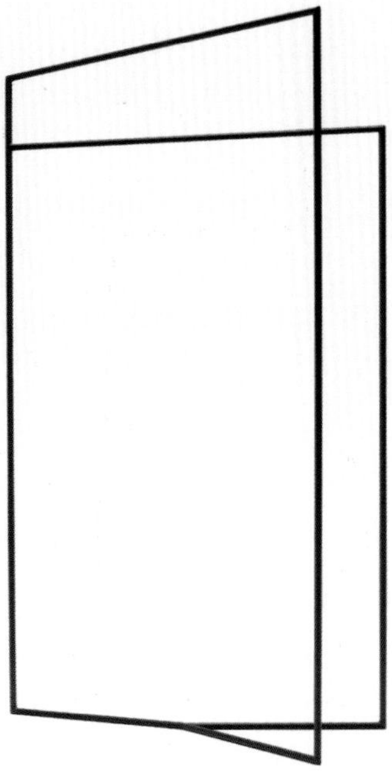

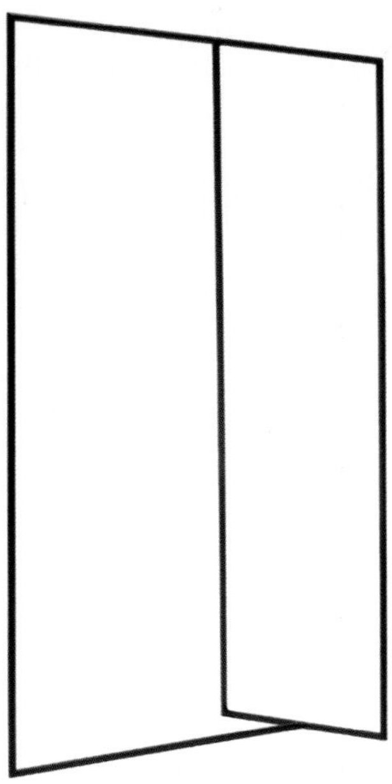

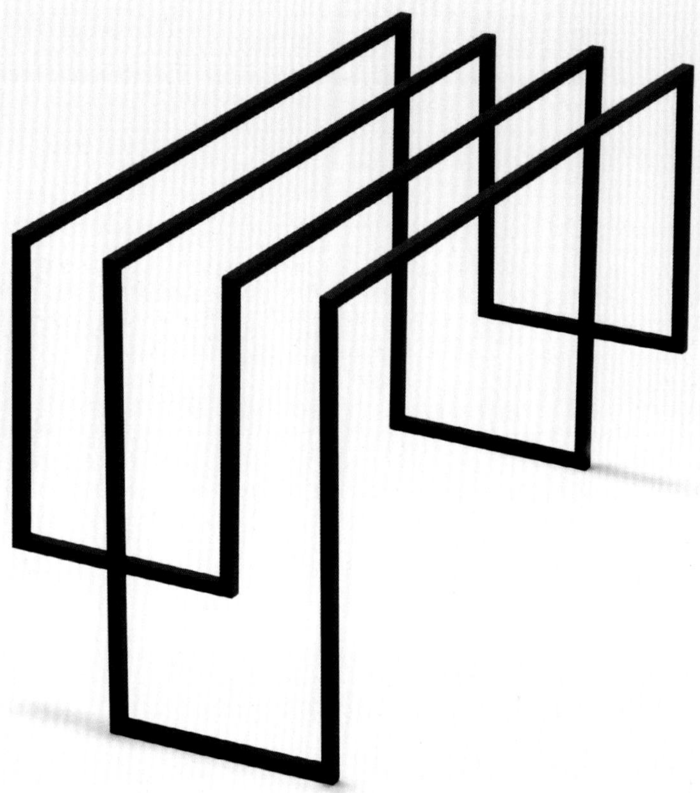

179

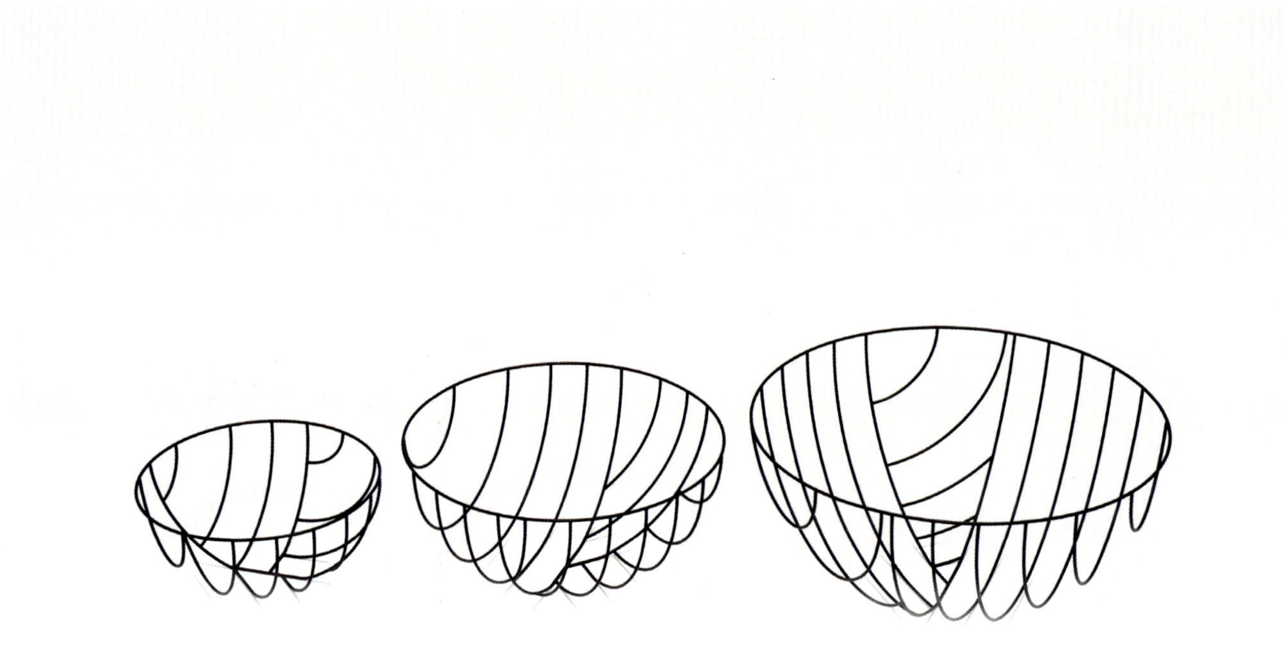

180

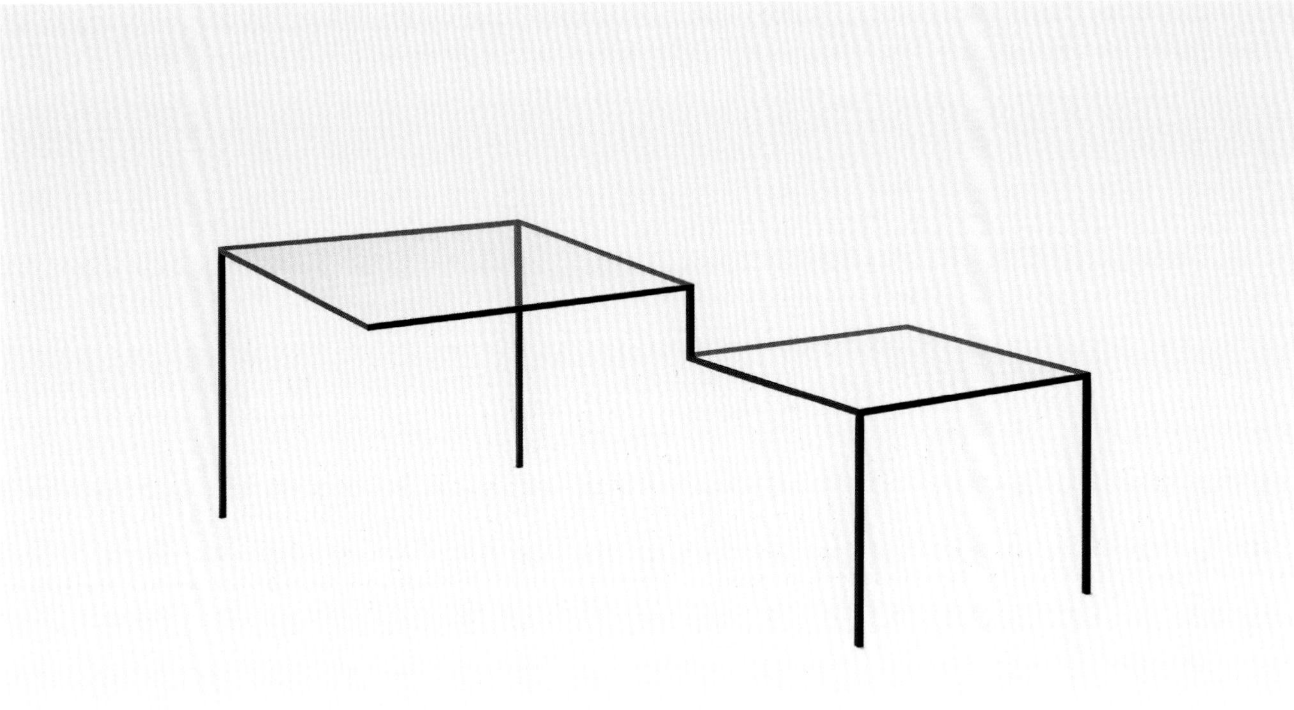

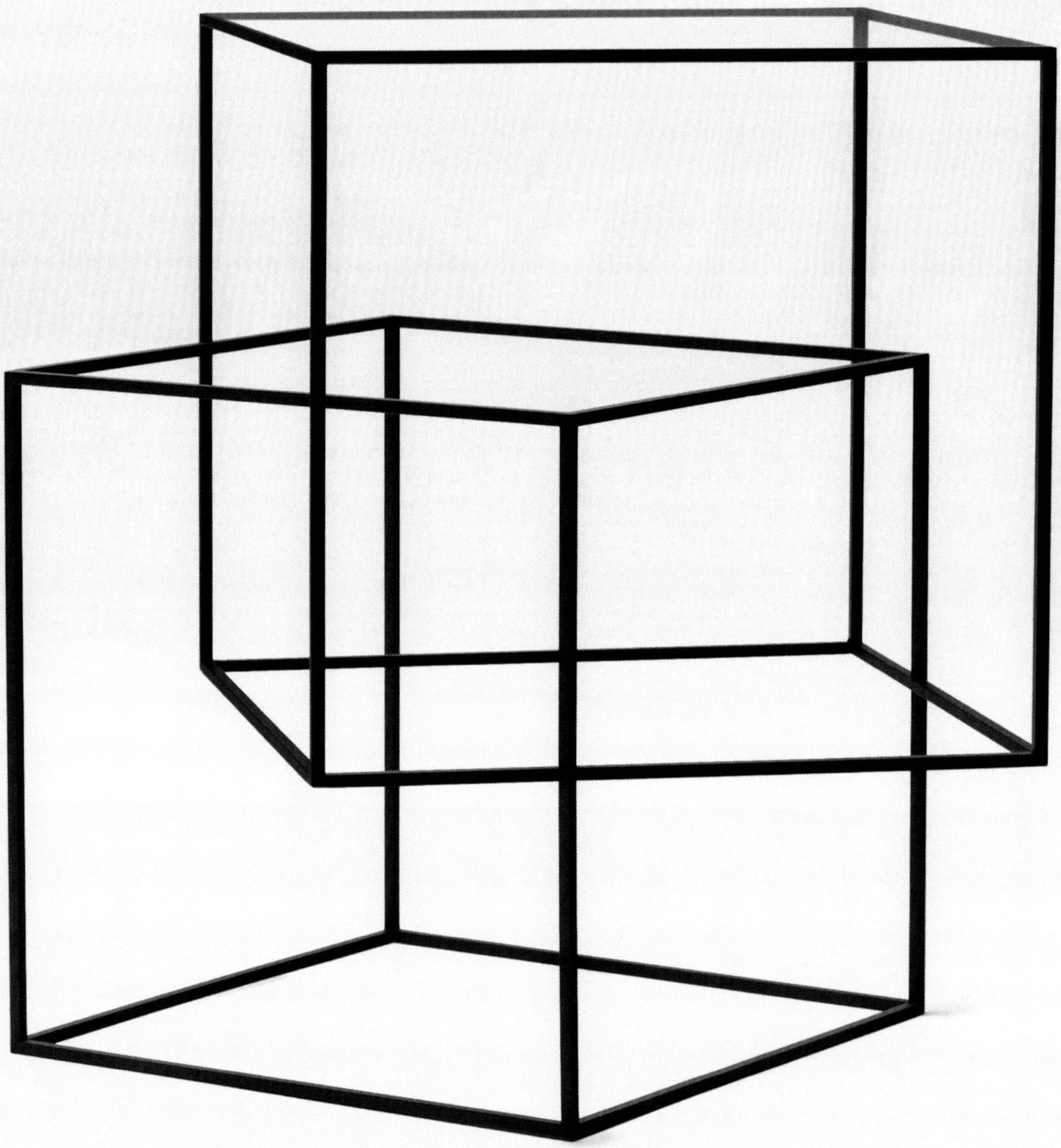

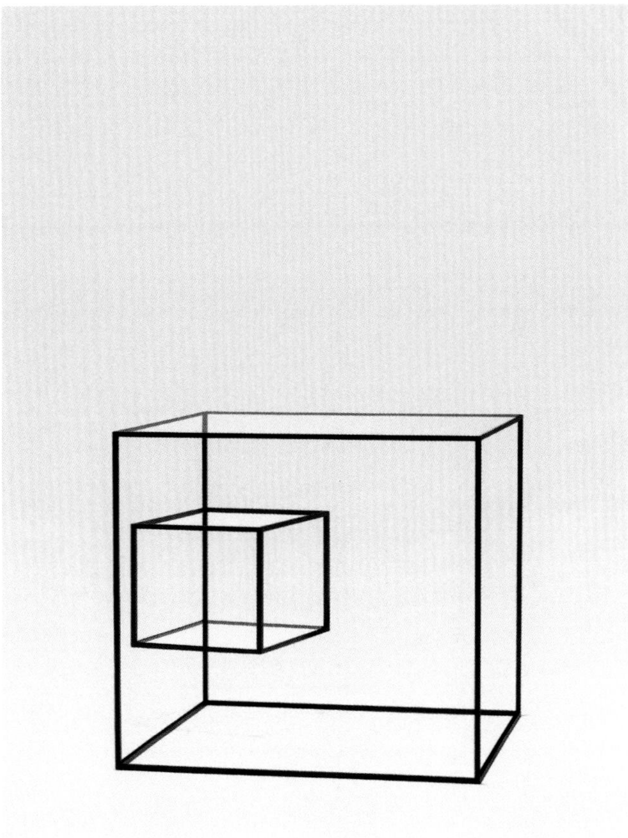

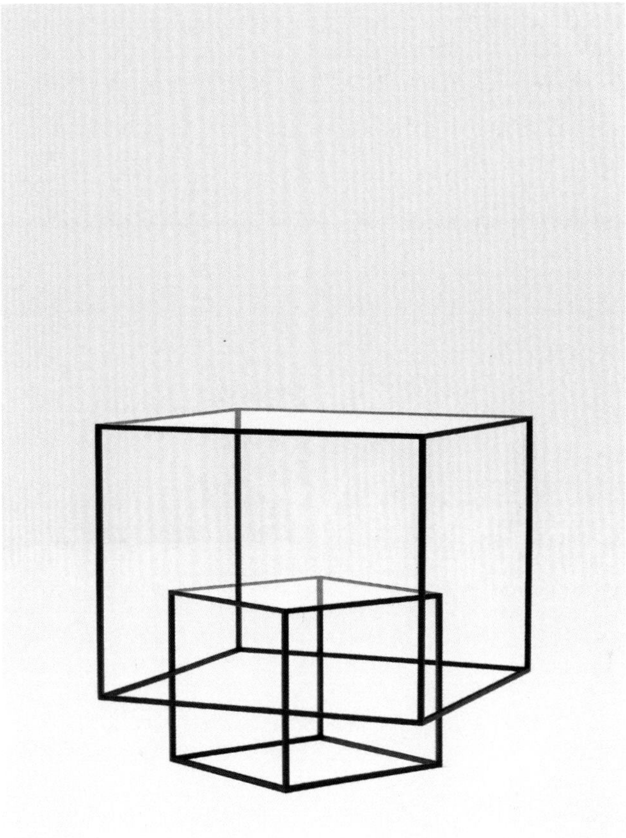

184

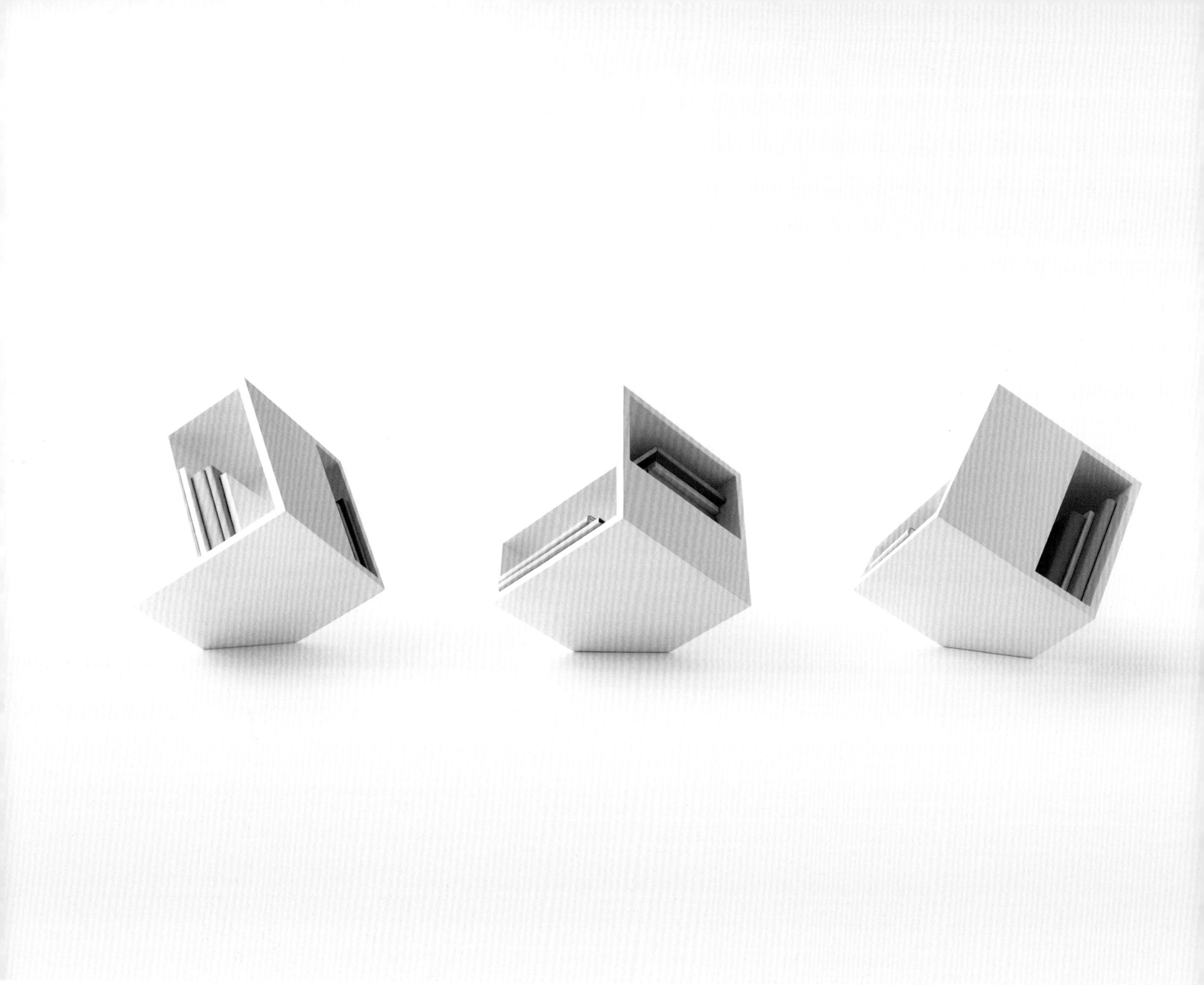

Cube
Arketipo

A cubic magazine rack that appears to balance on one corner. Unlike an ordinary box, which has only one entry point, Cube opens in three directions and is divided into three complementary spaces inside. Its distinctive tilt makes it easy to slide magazines in and out – and proves more supportive than a traditional magazine rack, making it suitable for any size of periodical.

Pyggy-Bank

Isetan

The name 'piggy bank' dates back to medieval Europe, when coins were saved in household jars made of an unglazed reddish clay named 'pygg'. This design takes inspiration from history, offering users a pig-snouted bottle or jar made of unglazed clay in which to place their loose change. It exists somewhere between 'pygg' and 'pig'.

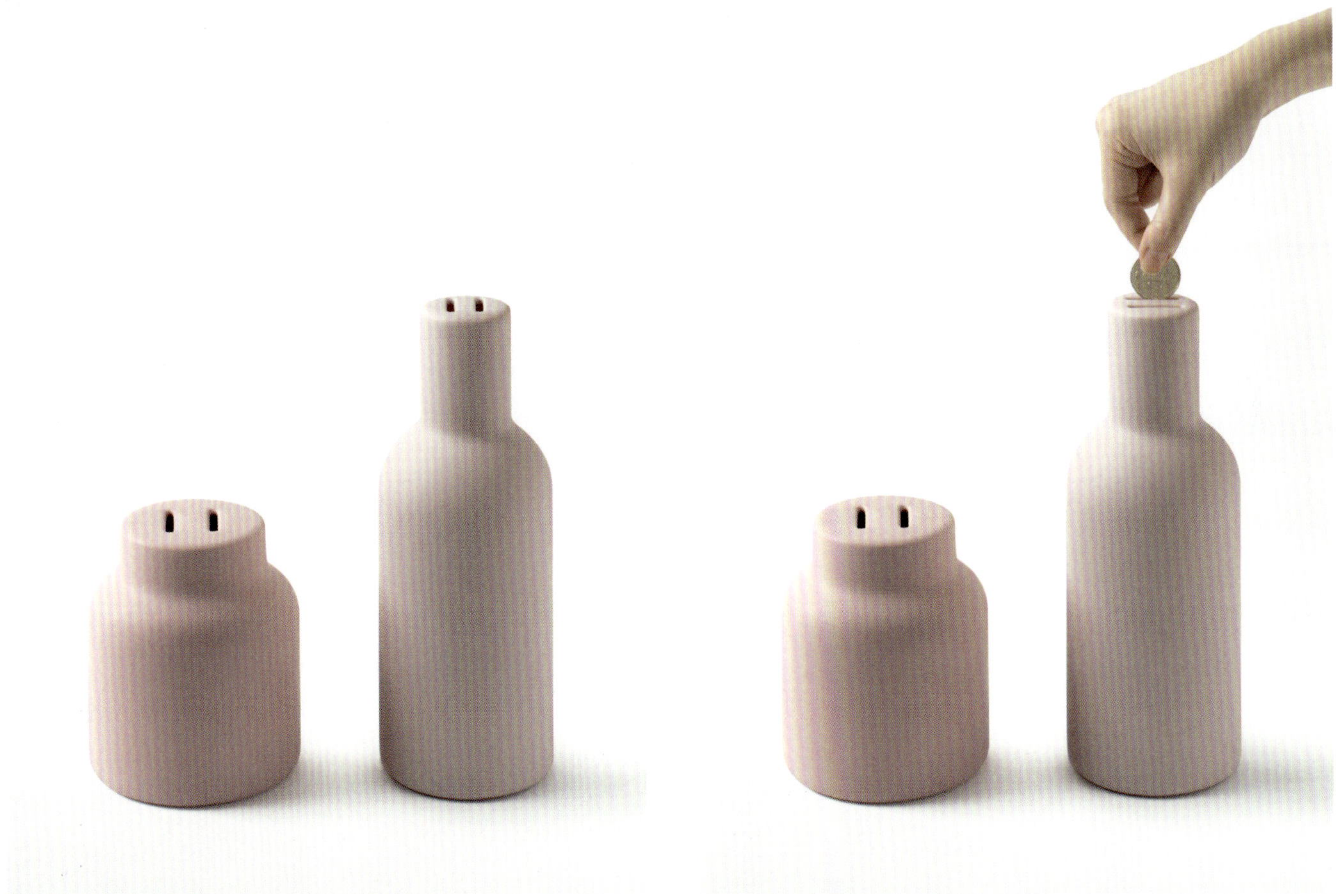

Tokyo Baby Cafe

Tokyo Baby Cafe

A 'parent and child cafe' in Tokyo's Omotesando district, designed to allow parents to enjoy being out with small children without worrying about those around them. The cafe is stocked with picture books and toys, and includes a playroom, private rooms and separate spaces for nursing and baby-changing. Wide aisles facilitate movement with a buggy, and light switches and door handles are placed high up on the walls in order to keep children from using them.

The space is designed to be enjoyed by two very different sizes of user – so the interior plays on this variation in scale. These users also see the world through different eyes. Take a table, for example: adults live their lives aware of tabletops, and the things placed on them. But children see the table's underside. Its legs can look like pillars to them, and from underneath, the tabletop is like a roof. The cafe's gigantic and miniature furnishings both take advantage of these two perspectives. A nursing sofa becomes a playroom when blown up to a massive scale, and a nappy-changing table when shrunk to minuscule proportions. Oversized windows pair with undersized ones, and big lightbulbs with small ones. The floorboards vary in size across the room, and the undersides of tables, where parents' eyes don't reach, hide pictures of parent and baby animals. In fact, illustrations of such parents and children can also be found all around the cafe, hidden behind tables and shelves, ready for discovery by their parent-and-child visitors.

2011

Non-Slip Birdhouse
Alcantara

A design for the 2011 *Can You Imagine?* exhibition at the National Museum of XXI Century Arts (MAXXI) in Rome, for which ten designers created new works using Alcantara – a microfibre material combining polyester and polyurethane. One of Alcantara's many unique material qualities is its non-slip surface; automobile manufacturers use it to upholster car seats, helping to ensure driver stability. To take advantage of this particular characteristic, a series of birdhouses was created, which, along with the slanting wall on which they were hung, were finished with Alcantara. Due to the material's suede-like texture, the birdhouses stayed on the wall through friction alone, and required neither glue nor screws.

Scatter Shelf

Friedman Benda, New York

The Scatter Shelf is composed of 5-millimetre (3/16-inch) black acrylic shelves in three grid forms, each of which is slightly offset from the others. The resulting shelving unit is not only structurally stable but when viewed at an angle, the glossy surfaces create a series of reflections, making the opaque acrylic appear to be transparent. The diffused reflections caused by the surfaces' sheen and staggered form also dissects and scatters the view behind the shelving unit, creating a kaleidoscopic effect.

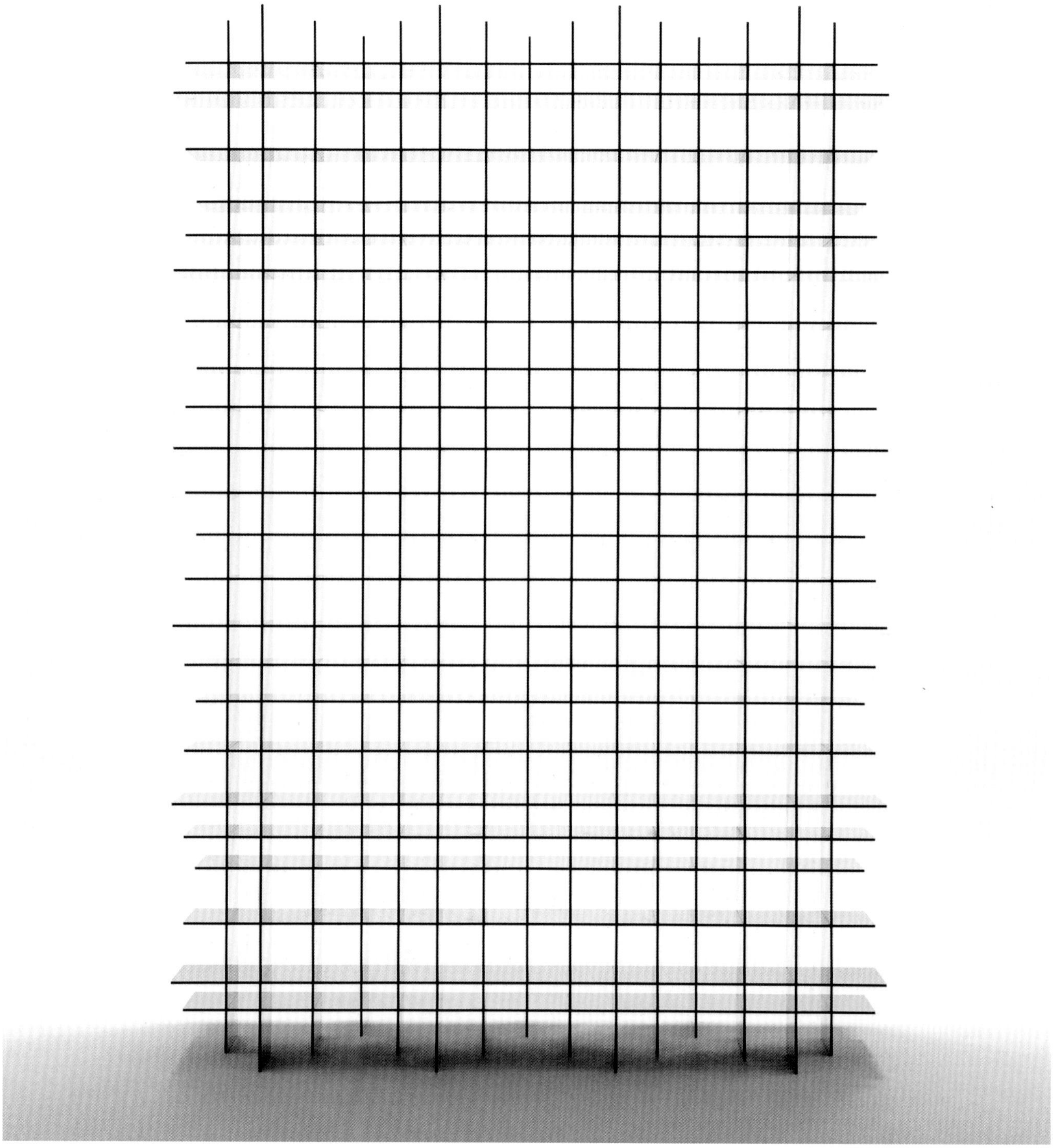

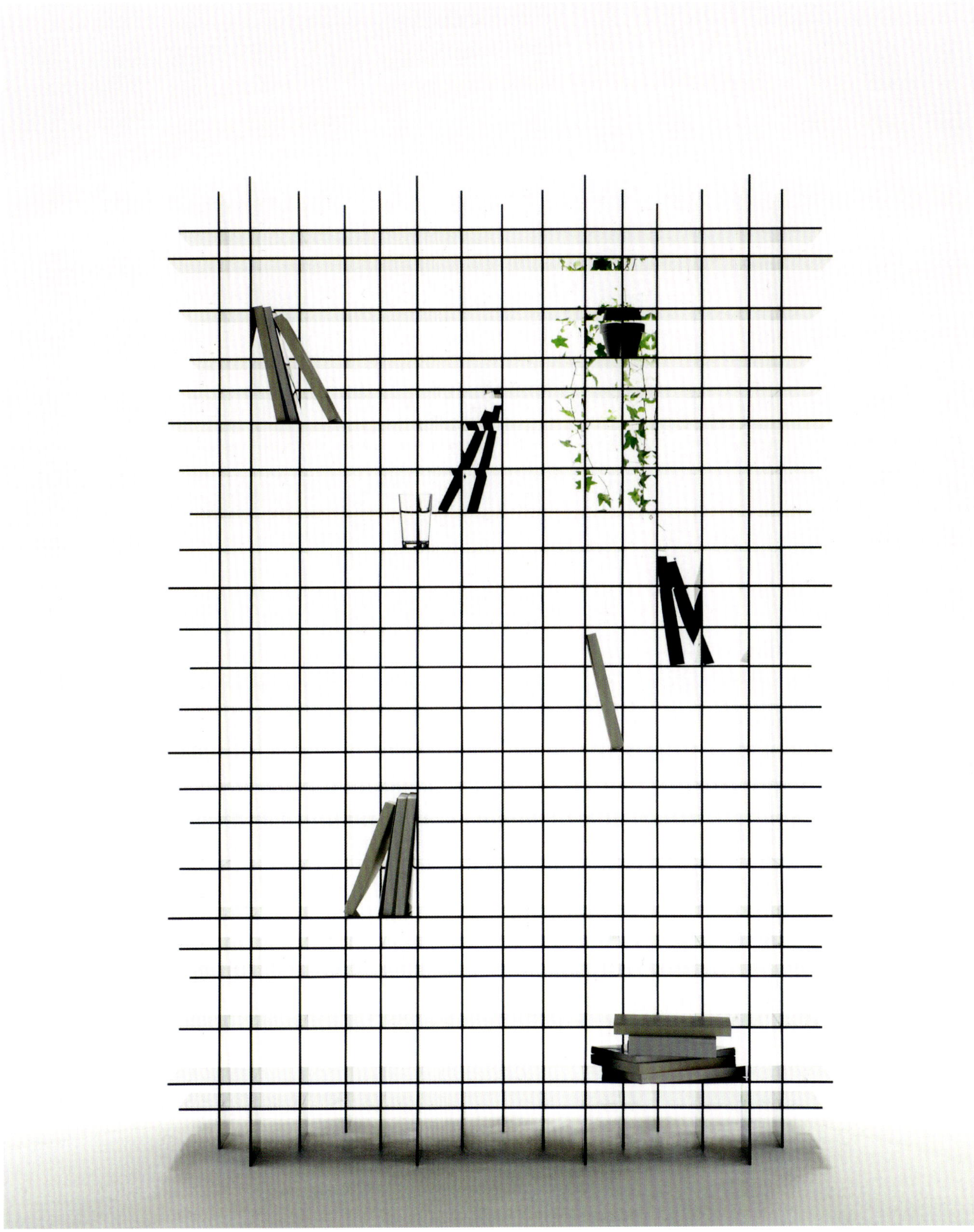

Dancing Squares
Art Stage Singapore

A series of objects in which square planes are assembled to create a sense of motion. The twist of the Dancing Square stool's 'legs' makes it look from some angles as if it is striding forward in a literal dance step. Near-spherical Dancing Square lamps roll about on the floor but are stable thanks to the flat surfaces of their constituent planes, and cast light in many directions. The Dancing Square table leans as though falling away, but maintains its function as a table, and makes objects placed on it seem to sink into its surface. All these different movements create an overlap between the states of balance and unbalance, as though the planes themselves are dancing.

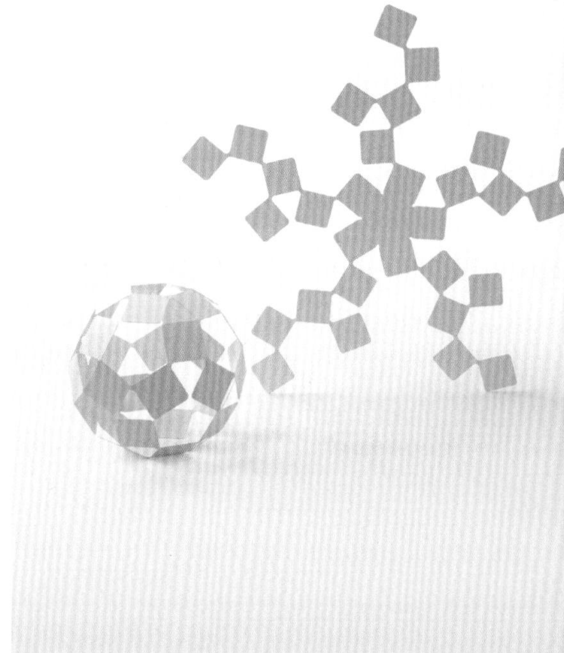

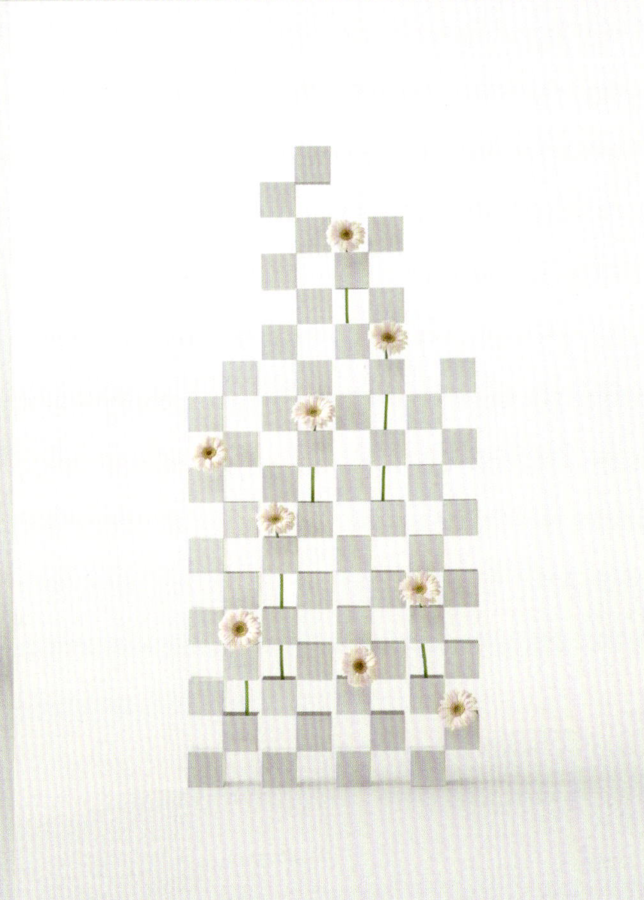

Thin Black Lines + Dancing Squares

National Taiwan Craft Research and Development Institute

A solo show held at the National Taiwan Craft Research and Development Institute, featuring, for maximum visual contrast, two collections – 2010's Thin Black Lines, and Dancing Squares from 2011. The former is a collection of furniture in 'still black', a term coined to express the static nature of these wiry, tubular-steel objects. In order to emphasize (and challenge) these specific qualities, the idea was to incorporate a theme of 'active black on white' for their particular exhibition space, which was represented by lines drawn on the floor that flowed like river currents around the display stands.

By contrast, Dancing Squares is a collection based on the concept of 'active white' – referring to its all-white surfaces, which are often tilted at kinetic angles – so its space was designed around the theme of 'still white on black'. Using a fisheye lens-like effect to alter the perspective of the space, a room-sized sketch was affixed to the curving walls and the floor, creating an impression similar to viewing the world through a tiny water drop.

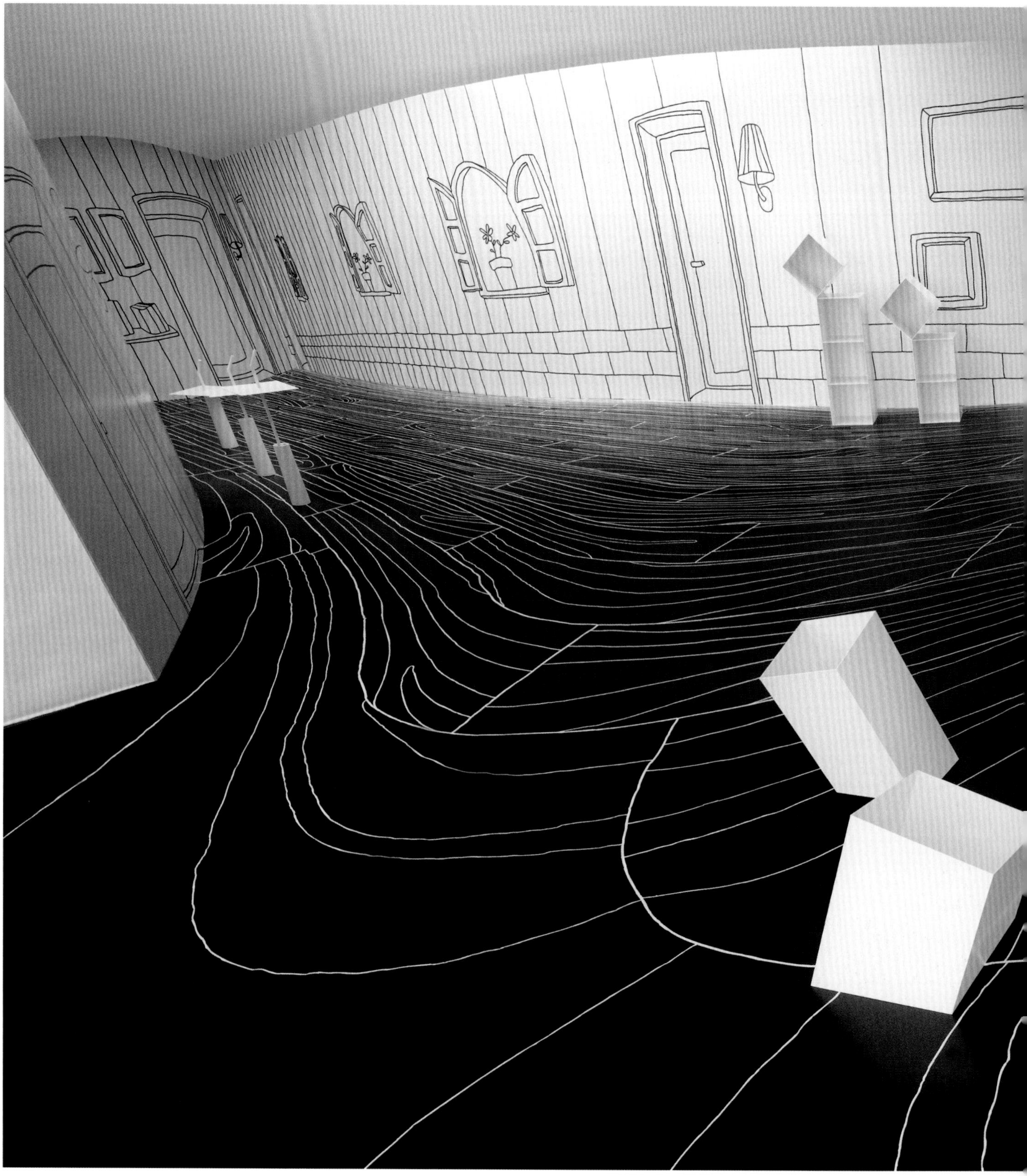

Forest-Spoon
Ichibanya

Spoons are tools for eating and are usually designed to serve this purpose alone. Forest-Spoon's design, however, is intended to appeal and excite even when the utensils are not in use. When a Forest-Spoon is presented alone, after it has been washed and is draining in the cutlery stand, for example, it looks like a single tree – but when in a group, it creates a 'forest landscape' in the kitchen.

Par
Materia

A salt and pepper set in which cork is used as a base as well as in its traditional role as a bottle stopper. Since the vessels can be upended and stored on the flat tops of their oversized stoppers, the design of Par subverts cork's subordinate relationship in the condiment set without denying its original function.

Parte

Materia

Influenced by cork's famous lightness and malleability as a material, Parte is a fruit bowl constructed out of two solid-cork parts. The twin halves are held together by invisible magnets, and can be connected along either their shallow or tall section profiles, allowing users to change the bowl's proportions and dimensions. The result is a design in which two different whole forms can be created by two identical halves.

Growing Vases
Lasvit

A chandelier that suggests both incompleteness and growth. By incorporating the metal pipes used by glassblowers into the design, still attached to the objects that they were making, Growing Vases displays a strong connection to the artisanal, handcrafted traditions used in Bohemian glassmaking. Furthering the concept of growth, the pipes were made to appear like flowers and branches, and the glass a series of vases, which reversed the typical arrangement of flowers blooming in vases to vases blooming from flowers.

Yii Exhibition

National Taiwan Craft Research and Development Institute

An exhibition-space design for *Yii*, commissioned by the National Taiwan Craft Research and Development Institute. Fifteen designs were exhibited at Milan's Triennale Design Museum, which had been generated by fusing elements of Taiwanese daily life with traditional crafts.

The installation comprised a 'forest' of 190 transparent vinyl balloons, 10 metres (33 feet) in height, which recreated the feeling of one of Taiwan's many bamboo groves. It was an appropriate spatial expression for the project's aims to bring together contemporary society and nature, and industry and craft.

Puma House Tokyo
Puma

Puma House Tokyo, located in the Japanese capital's Aoyama design district, combines the company's press room and event suite into one space for the first time. It is a multipurpose space that can be used for exhibitions, fittings, product launches and other media events – and is available for hire.

Dark-timber 'staircases' climb around raw-concrete columns like vines. But these staircases are not for people to use; rather, they function as display stands for Puma's trainers and as a compositional element that gives the space a dynamic character. Nonetheless, the resulting effect is a strong reminder that we exercise our bodies daily going up and down stairs – and also has a visual connection with stadium stairs and podiums, to reference the company's important relationship with the world of sport. The stairs bring a sense of movement to the interior, enabling a three-dimensional product display that uses its space to express physical activity and allows visitors to fully experience Puma's brand.

GIVE
75%

Transparent Chair

Milan Design Week

A chair whose body is made from a thin film of polyurethane. Thanks to its high elasticity and ability to return to its original state, this transparent material is commonly used for packaging precision instruments and delicate products that might be susceptible to vibrations and shock. Looking at the chair, it seems to consist of nothing but a backrest and armrests – with no upholstery or cushioning to speak of. Yet, the near-invisible film wraps and supports the body like a hammock, providing a light, floating feeling for the sitter.

Transparent Lamp

Milan Design Week

A pendant lamp whose protective film shade is translucent when seen head-on, and transparent when viewed at an angle. The light source is placed at the centre of the ring of film, creating a lampshade that, owing to its circular shape, mutes the light at its centre by its translucency, regardless of the angle of view. The outer area, however, remains transparent.

Transparent Frame

Milan Design Week

A picture frame created from artificial Ulexite – a mineral that, even in its natural state, has the fibre-optic-like ability to transmit images of objects adjacent to it without diminishing their quality. Artificial Ulexite is not only transparent but also makes objects seen through it appear to float on its surface, as though depthless. This property of the material is used to 'transfer' part of the image supposedly contained inside the frame on to its surface. This creates an ambiguous boundary between image and frame, changing the way in which we normally view images.

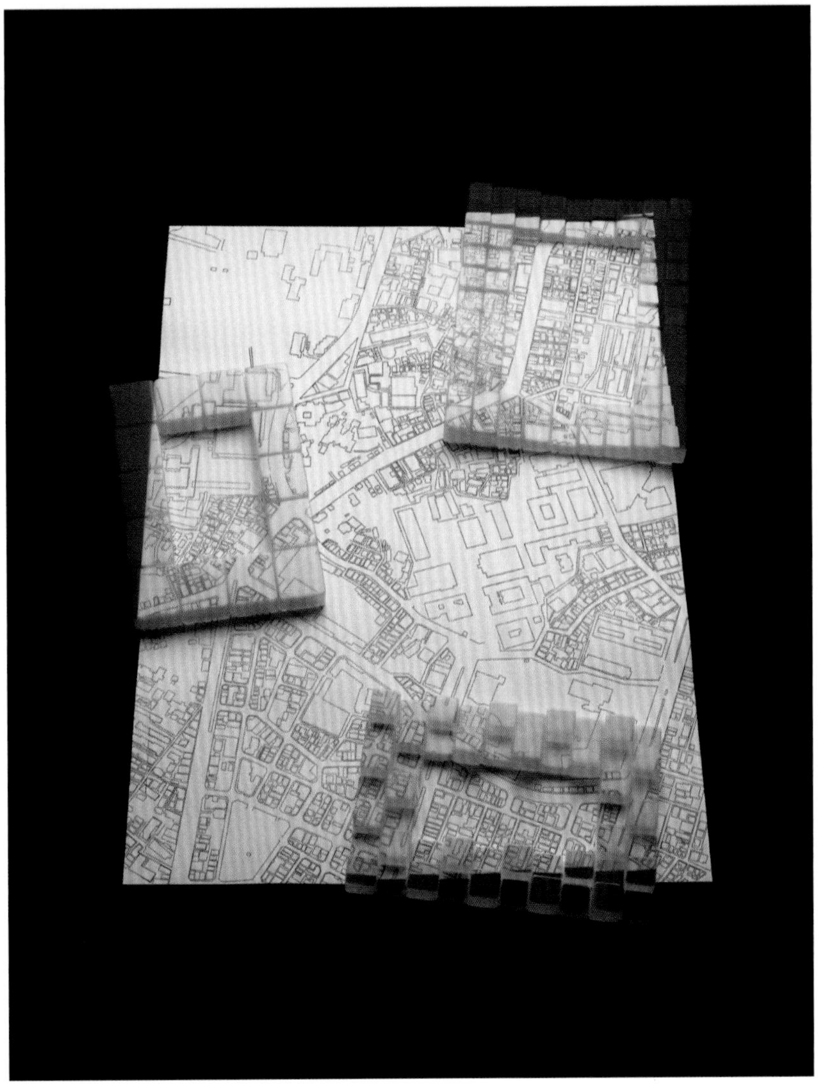

Transparent Table

Milan Design Week

To achieve the effect of transparent wood, clear acrylic was cast in strongly grained timber formwork and the resulting pieces assembled into a table. The planks' butt ends were faithfully reproduced and the edges bevelled like floorboards. The grains and the dimensions of the wood used for the legs were matched to those of the tabletop, unifying the piece.

The collection's two table variations generate opposing optical effects: on first sight, the black version is made of wood, but a longer look reveals its transparency; the clear version, meanwhile, appears initially to be transparent, only revealing its wooden characteristics on closer inspection.

50 Projects_25 Objects + 25 Spaces

Axis Gallery, Tokyo

A solo exhibition of twenty-five products and twenty-five interiors, selected from the myriad projects in which nendo has been involved since its establishment. Transparent, hollow polycarbonate sheets were stacked to make a 'wall' in front of each item displayed. This created a pixelated effect and meant that visitors had to go around the back of each wall to see the work. The degree of 'pixelation' varied with the angle at which visitors looked through the polycarbonate. Some walls overlapped, creating an environment in which the view changed like that seen through a kaleidoscope.

Kotoli

Ruinart

A picnic box designed for Ruinart, the world's oldest champagne house. The company was interested in offering a new perspective on urban living, celebrating all the city has to offer through picnicking rather than attempting some rural escape.

Kotoli references the way in which birds swoop down to perch on a telephone wire. The stemless champagne flutes cannot stand on their own. Instead, they can be slipped casually into the breast pocket of a suit jacket, or the silver ring attachment allows them to 'perch' for a rest on the bottle or the picnic box. This versatile arrangement enables urban picnickers to enjoy champagne anywhere, even far from a tabletop.

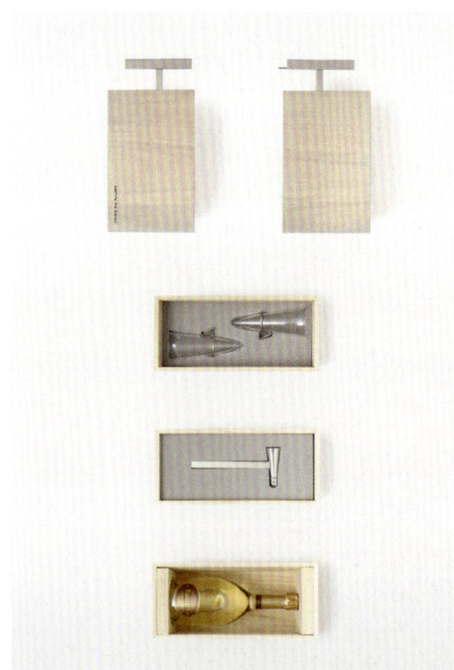

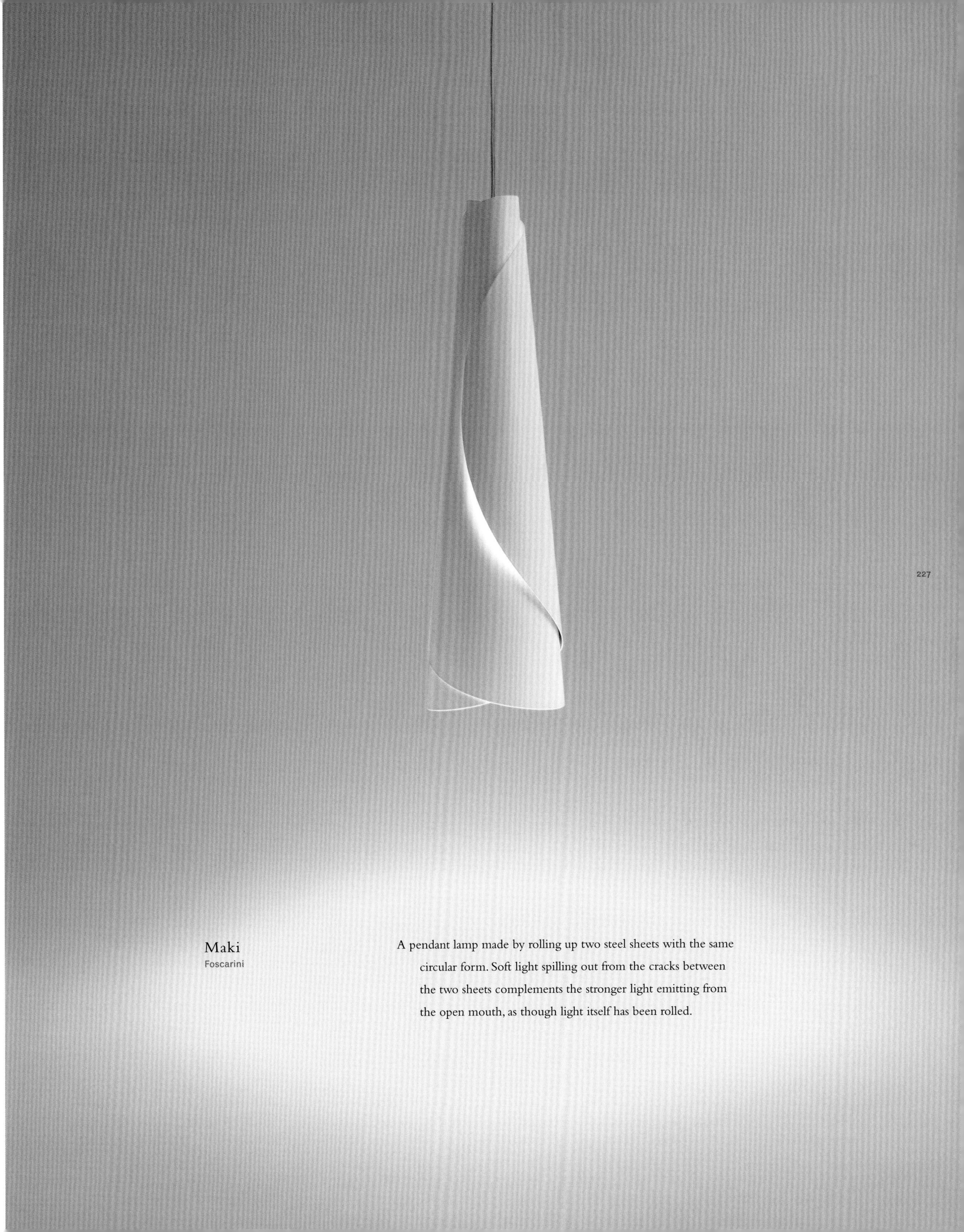

Maki

Foscarini

A pendant lamp made by rolling up two steel sheets with the same circular form. Soft light spilling out from the cracks between the two sheets complements the stronger light emitting from the open mouth, as though light itself has been rolled.

Data Clip

Elecom

A USB stick that mimics the form of an ordinary paper clip. Like this everyday stationery item, it can attach to a stack of documents, business cards or memos for handing data to a colleague or friend. However, it can also connect to other Data Clips for easy storage. Its design aims at a new, more integrated, relationship between daily life and the digital data that nowadays forms such a big part of it.

229

Oppopet

Elecom

A wireless, optical mouse whose USB receiver can take the form of different animal tails (*oppo* means 'tail' in Japanese). When transporting the mouse, users can plug the *oppo* into the body and carry it as though picking up a small animal. When in use, the mouse takes on a simple, discrete shape and the tail's USB end plugs into the computer, extending casually as though a small pet is hiding inside the machine.

Established & Sons at LDF
Established & Sons

An installation for British home-goods company Established & Sons' exhibition during the 2011 London Design Festival, presented in the company's showroom in east London. The collection's theme was 'My London', and, in response, 15,000 maps of London neighbourhoods were printed onto tracing paper and arranged on the walls, creating four spaces that recollected that particular atmospheric phenomenon, the famous London fog – particularly the 'Great Smog' of 1952. The installation – which was intended to evoke intimate, individual feelings towards the capital among its viewers – became a physical, spatial expression of the dynamic, always ambiguous, face of the city.

Visible Structures

High Museum of Art, Atlanta, Georgia

A collection that explores the new expressive possibilities of carbon fibre by using it in a 'supporting role', allowing the emphasis to be placed on the main material. Comprising stools, chairs and tables, the furniture series was constructed from foam core and then reinforced with industrial-strength carbon tape.

Carbon fibre bends when used on its own, but when 'sandwiching' weaker materials like polystyrene, the other material's compressive strength works to suppress its pliancy – creating a strong, light composite material. Adhering the carbon fibres to the surface of the furniture like tape or a sticker achieves the strength necessary for it to fulfil its various functions. This new method of making overturns the traditional composition of 'internal structure and external cladding', as well as the basic manufacturing process in which the structure is first formed then finished – common to both architecture and furniture – ultimately reorganizing the relationship between structural form and surface material.

The idea initially grew out of carbon fibre's use as a supporting building material after the Hanshin (or 'Kobe') Earthquake of 1995 in southern Japan. Wrapped around pillars or bridge girders, it provided additional seismic resistance.

Hirata No Boshi

Akio Hirata

An exhibition design for the first major retrospective of internationally known milliner Akio Hirata's seventy years of work. The mass-produced, non-woven fabric hats created for the space were the antithesis of Hirata's carefully handmade objects, and brought the two into sharp relief through dramatic contrast. Hirata himself oversaw the shape of these hats, which floated and streamed through the exhibition like ghosts or shells of the real hats exhibited. Some were stacked in order to serve as exhibition stands; others formed walls or ceilings – or functioned as diffusers to scatter light through the space. Blanketed in roughly 4,000 of these 'ghost hats', which formed a vast cloud, the exhibition space softly invited visitors inside. There, they found not clear-cut paths to follow but an environment in which they could wander and discover Hirata's creations as they liked, as a way of physically experiencing the creative freedom that underlies his work.

238

2012

Innerblow
Carpenters Workshop Gallery, Paris

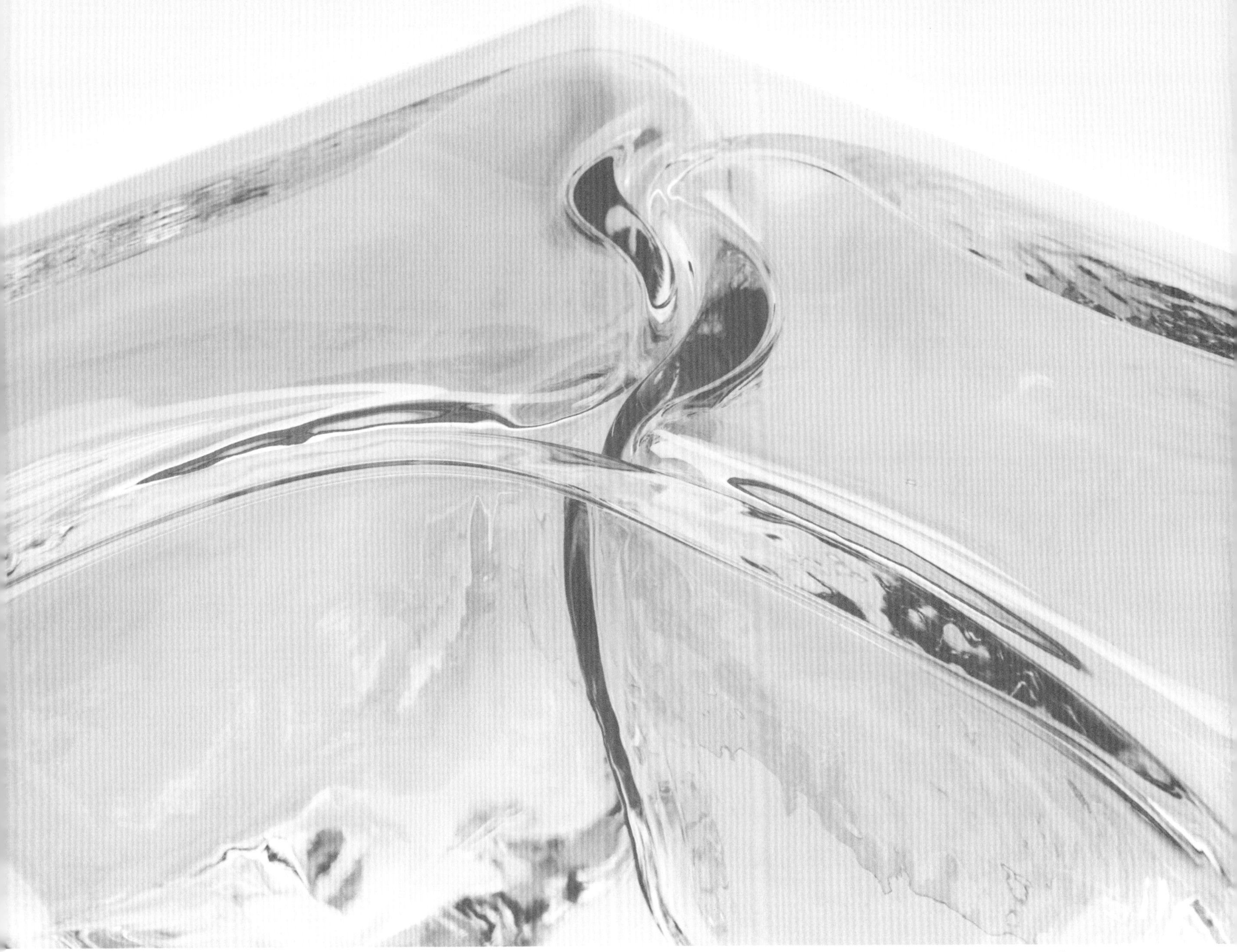

A furniture collection inspired by the way in which glassblowers use their breath to expand and shape lumps of molten glass into tangible forms. During the production process, glass was expanded into a square metal frame placed on the floor. The structure was then flipped over and its resulting flat surface, formed through pressure against the floor, used as a table top. Once upright, the mould remains in place, providing support for the table, as well as forming an integral part of the design. Each of the five tables in the collection is different. In one case, the metal form was raised slightly off the floor during blowing in order to allow the glass to swell out underneath, like the top of an inverted loaf of bread. For the bench design, several glassblowers blew into one mould simultaneously, creating a number of different sections that seamlessly moulded together.

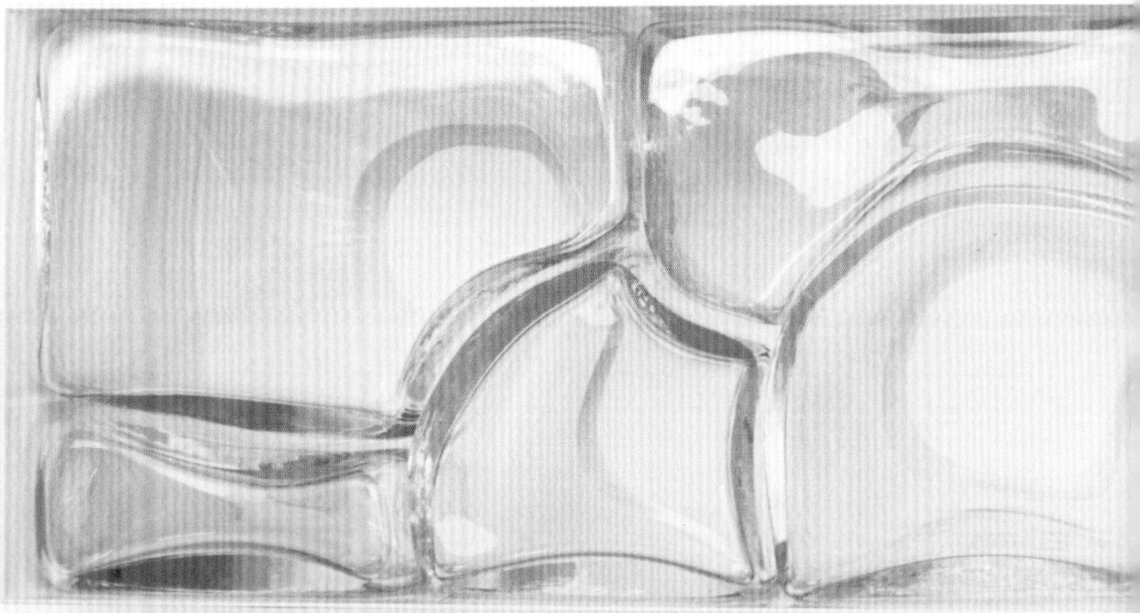

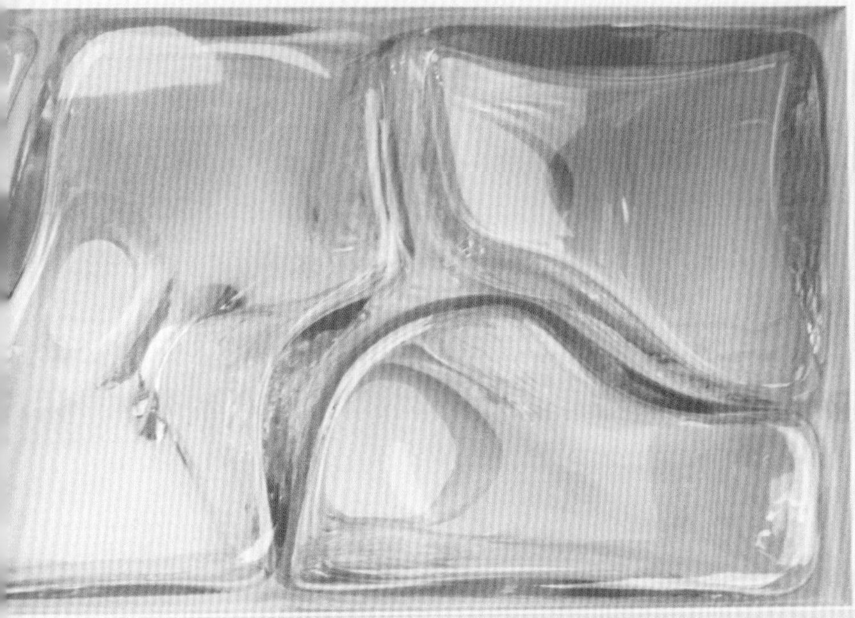

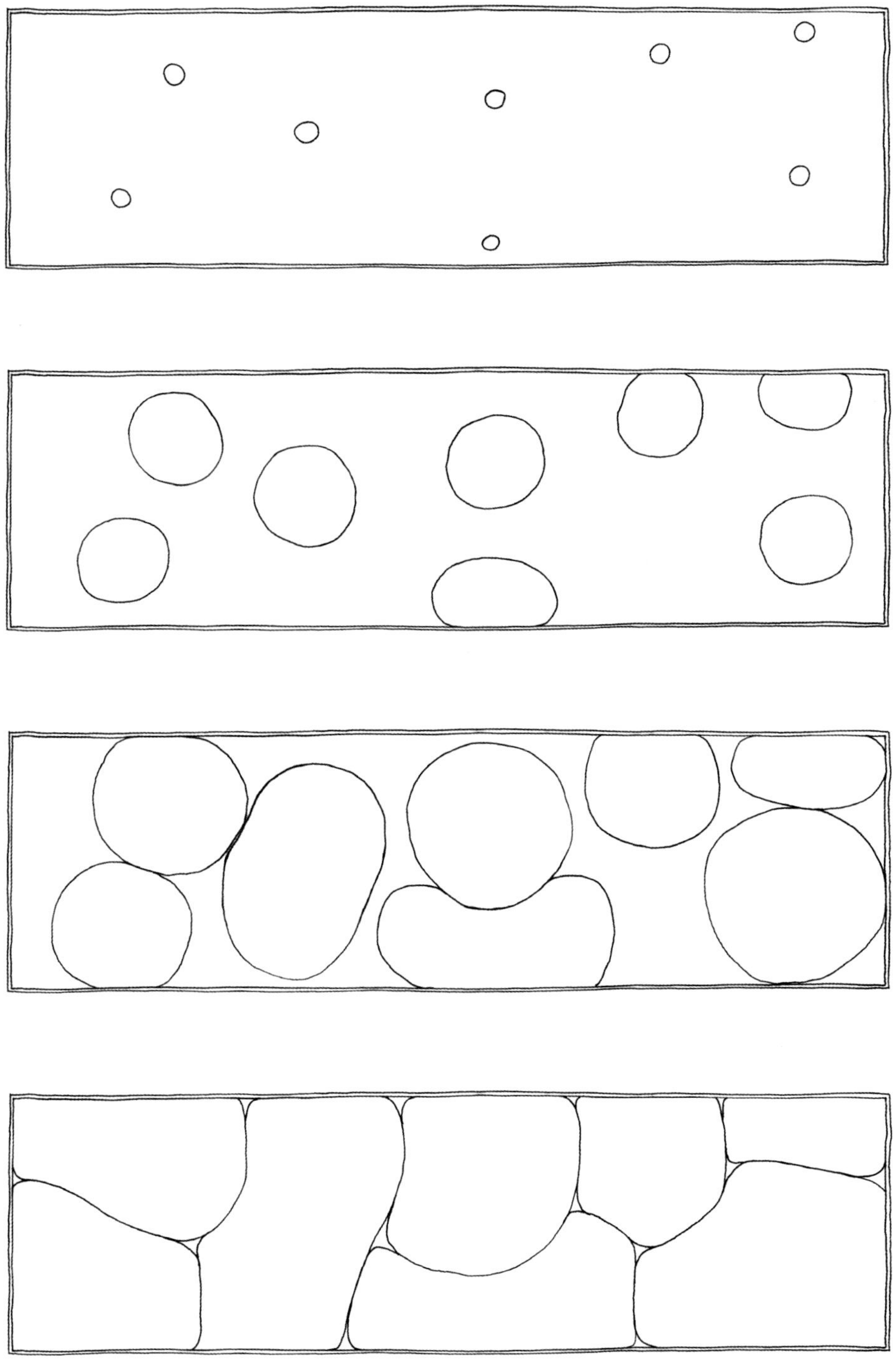

Overflow

Carpenters Workshop Gallery, Paris

Glass is formed when a free-flowing liquid cools down and hardens in place, creating a unique form. In this case, plate glass was placed in a frame with one section missing and was turned into molten glass by progressive heating, such that the material ran out of the missing section. By hardening the glass again at just that moment, the edge responds to surface tension, tending to minimize its surface area through the attraction of the particles in the surface layer by the bulk of the liquid. This naturally creates a rounded profile, giving the impression of a table edged like a pool of water. The collection's twelve tables vary in the amount of glass that escapes the frame, and in the position of the overflow.

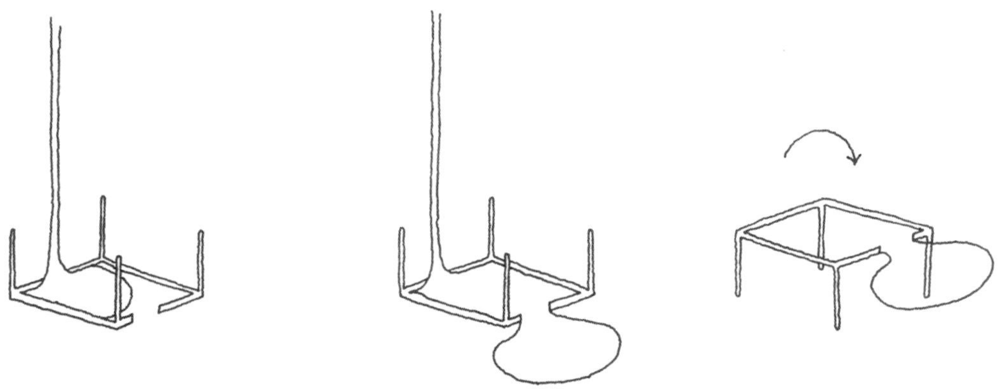

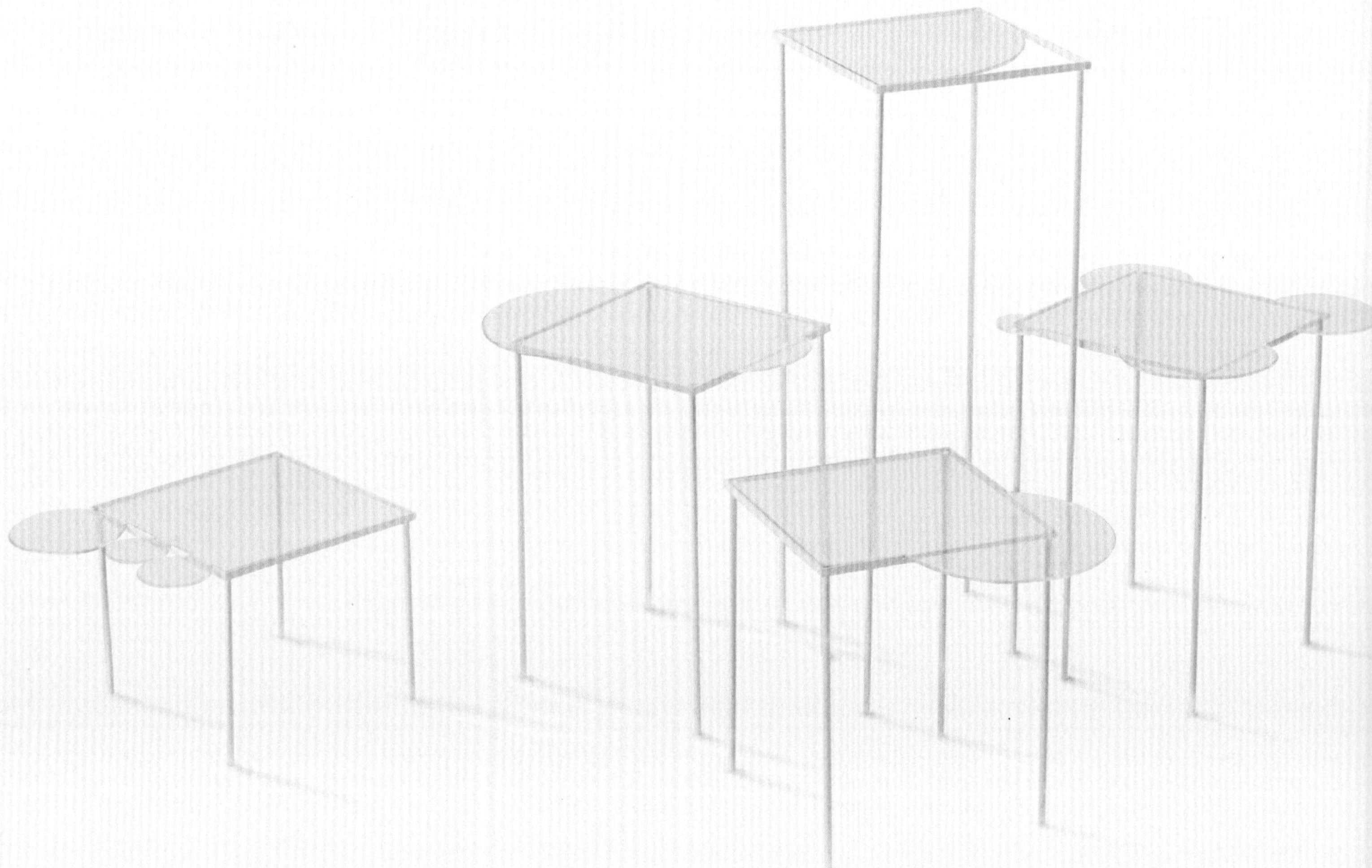

Farming-Net

Carpenters Workshop Gallery, Paris

A collection of sculptural objects made by heat-forming the sort of agricultural nets that are ordinarily used to protect fruit and vegetables from adverse weather and the threat of animals. The nets are stronger than muslin but more flexible than wire mesh. Here, their unconventional, sculptural properties evaded the traditional necessity of combining a structural material with a separate surface substance; instead, the nets were used to create a thin membrane that stands independently but also floats gently on a breeze. The Farming-Net flower vase and bowl seem almost to capture the air, like a *furoshiki* Japanese wrapping cloth. The lamp softly emits light in the manner of a traditional Japanese paper lantern. The table, with its multiple pleats, is reminiscent of the surface of Japanese *shibori* (tie-dyed) textiles and handmade paper.

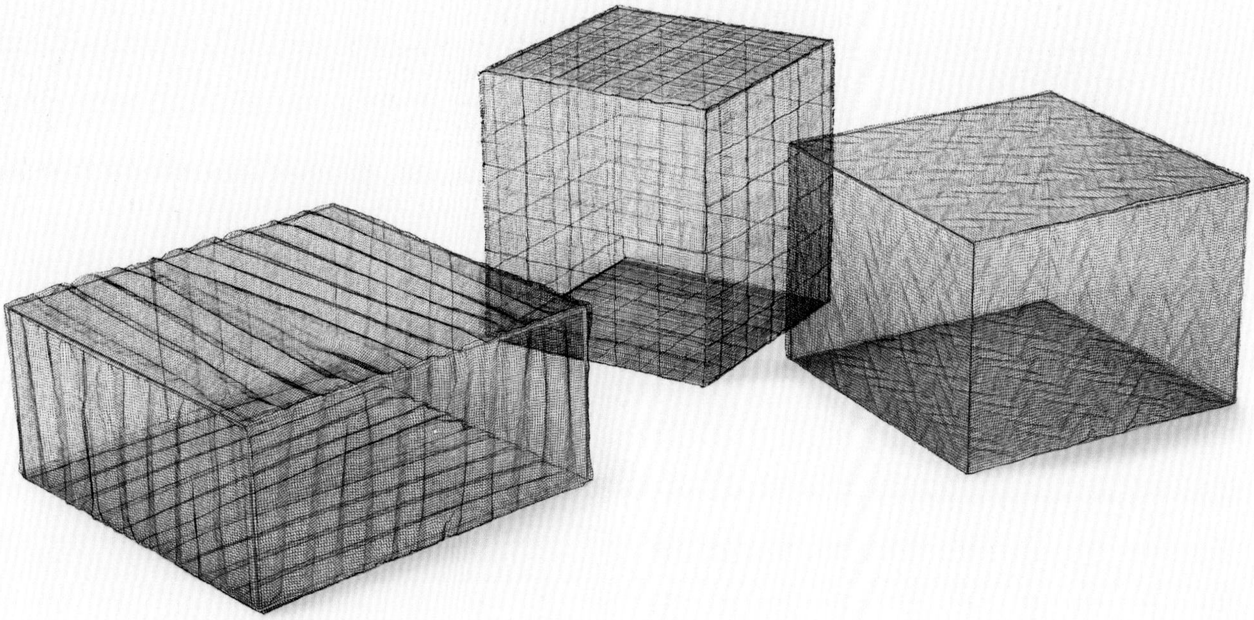

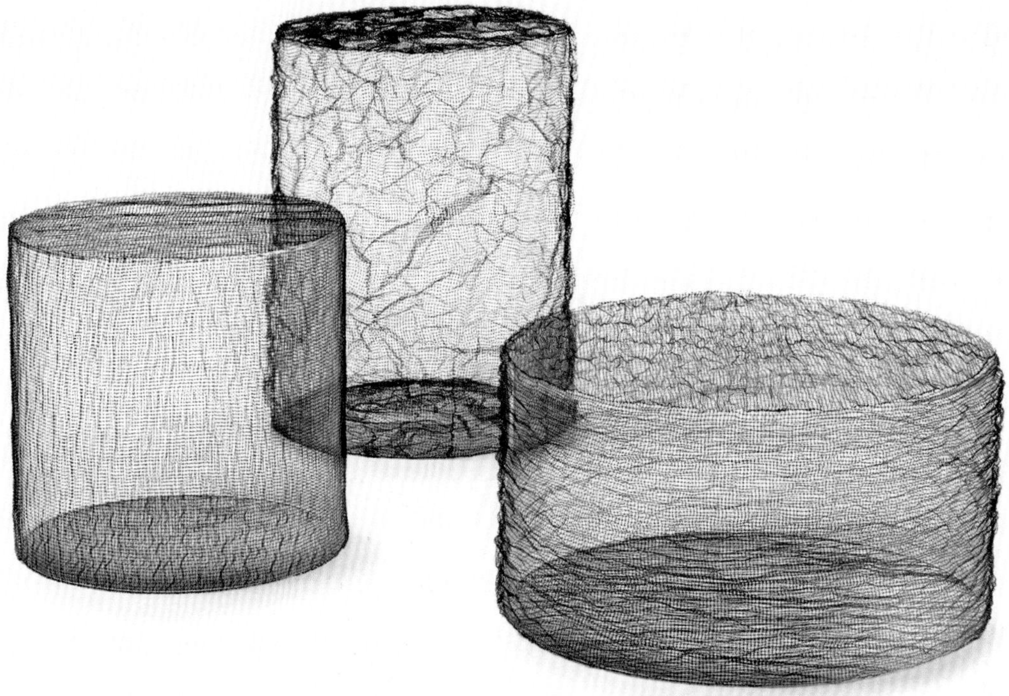

Top-Tea Set
1% Products

In this teapot-and-cup set, the thick wooden lids provide good insulation to keep the tea warm. The tapering section of the lid condenses steam into liquid and directs it back into the brew, rather than dripping down the sides of the pot. In a further, playful touch, the lids also function as spinning tops, and can be spun on the tabletop for amusement while drinking tea.

Stack-Sake Set

1% Products

A sake set comprising a pitcher and vessels of various sizes, which take the form of stacked cups. The different sizes correspond to different types of drinks: a small, single *choko* cup; a slightly larger 'two-cup'; a pile of four cups that creates a tumbler; and a stack of five that makes a pitcher. Because of their repetitive shape, they can be stacked together when not in use – and the one-cup *choko* can double as a lid for the pitcher. When drinking warm sake, the pitcher is heated in a pan of hot water and, in doing so, the uppermost *choko* cup/lid can be heated too.

Block-Vase

1% Products

A set of small, bud vases that can be stacked together like building blocks. The units come in four different sizes, specifically measured to fit neatly on top of one another. With careful stacking, and the removal of the upper vases' discreet base stoppers, the four units can accommodate a single, tall flower by passing its stem through the top vase and on through the unstoppered holes to the bottom one.

Twin Cup

1% Products

The coffee-filter holder section of Twin Cup mimics the shape of its companion coffee mug and saucer. Due to its shape, the filter holder is very stable and, of course, holds exactly the right amount to fill its coffee-mug 'twin'. After brewing, the saucer doubles as a tray for the used filter holder; when stored, the set stacks compactly by placing the holder between the saucer and cup.

Scissors

K%

A magazine rack formed of flat steel bars. The supports are arranged diagonally to create diamond patterns, but in slightly varying planes so that magazines can be held in the gaps created between them. It is hard to see the difference between the parts of the rack that support periodicals from below and those that keep them from sliding out at either the front or rear. As a result, the magazines look as though trapped in a giant spider's web.

Timber

K%

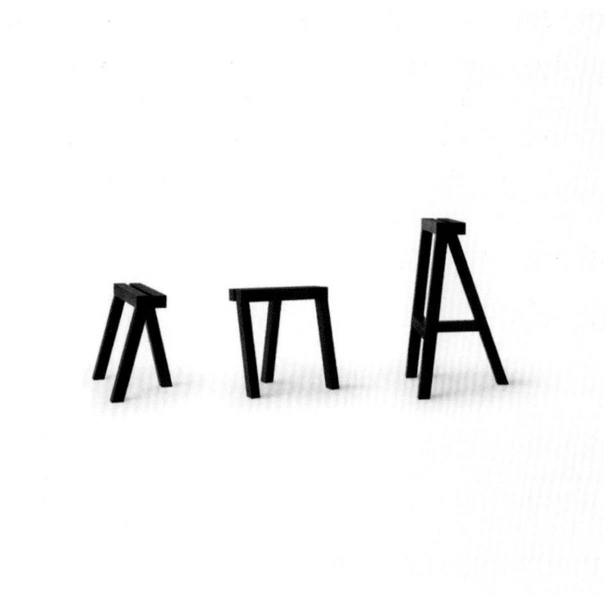

A stool created from three legs of timber, angled jauntily into a twin-timber cross-piece seat. In the larger version, the stool is also braced by a lower T-shape. This formation prompts a casual leaning posture in the sitter, whose legs spread forward for support, making it seem as if they are imitating the stool itself.

Rain

K%

A bookshelf that holds its contents with a series of 'lines' rather than with continuous surfaces. Rain is constructed of 8-mm (5/16-inch) -diameter steel rods – thirty of them deployed along the unit's X-axis, twenty-two along the Y-axis and twenty-eight along its Z-axis. Together, these hold the books in place.

Object Dependencies

Specimen Editions

Originally, furniture was structurally autonomous and served to hold, on its own, books, cups and other objects (including people) placed on or in it. This supporting function was one of the most important factors in determining the form of any given piece of furniture.

The Object Dependencies collection presents, by contrast, items of 'weak furniture' – among them, shelves, tables and cupboards – that cannot function until they are made structurally sound through the addition of objects. In addition to increasing stability, the accumulation of objects has other effects as well. It can change the angle of light emitted from its tilting, mousetrap-like lamp, held back by small weights laid on its restraining flex, or expand a bookshelf by sliding its weighted components across each other. The collection is an exploration of new forms created by resetting the once-unquestioned relationship between furniture and objects.

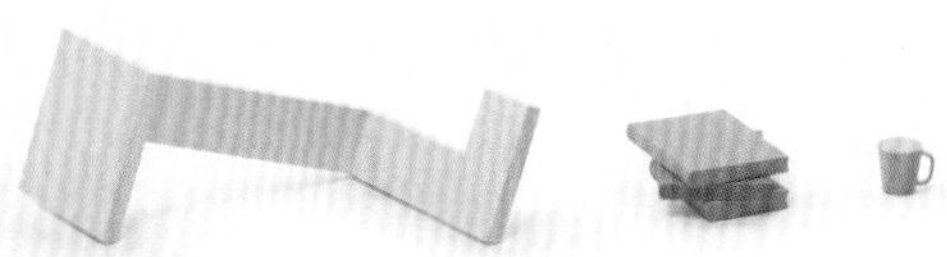

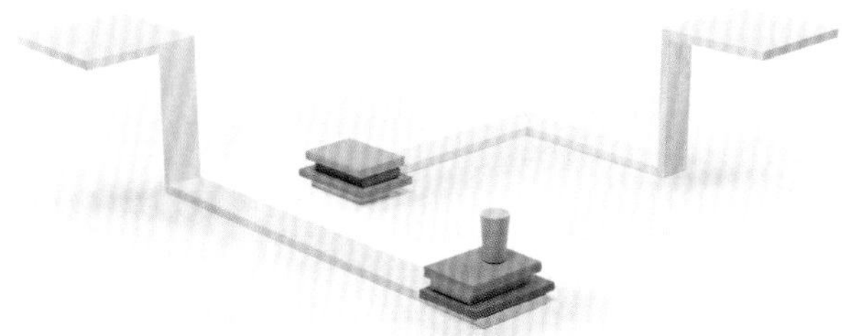

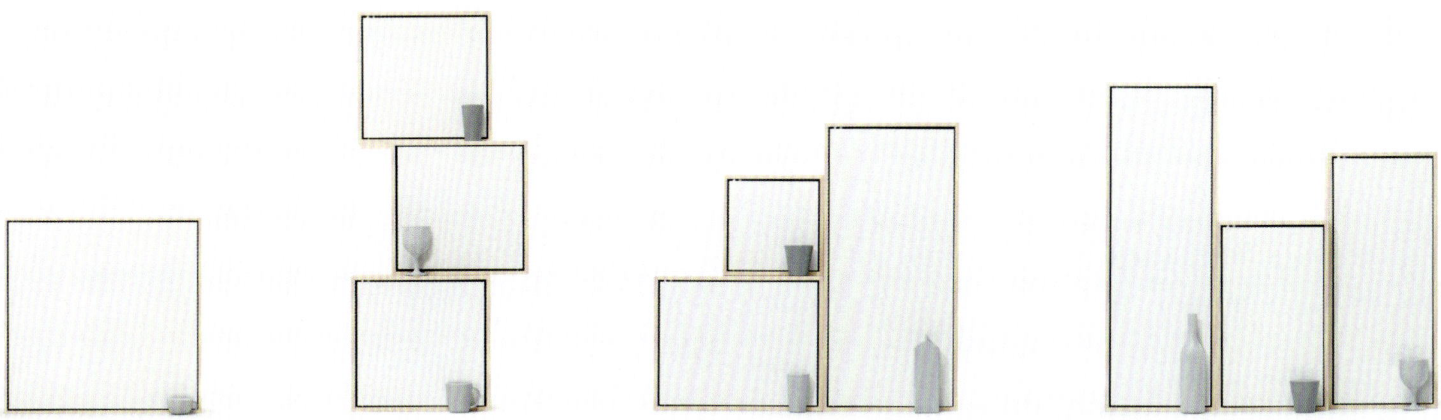

Lacquered Paper-Objects

Nilufar, Milan

A series of small containers created using a 3D printer that cuts, stacks and pastes sheets of paper together one by one. These stacks were then carved by hand in a workshop, sanded down and their surfaces finished with lacquer. As the lacquer is gently rubbed away, a mysterious texture, like wood grain, is revealed – thus resembling the raw material from which these objects originated.

Beige 1st Concept

Onward Kashiyama

The concept and name for women's clothing brand Beige derive from the natural colour of textile fibres; it offers women simple, functional designs that nonetheless showcase the charm of their materials.

Inspired by this philosophy of having respect for the natural state of things, the store design removed any unnecessary decoration and incorporated C-beam structural steelwork, creating a visual feature out of an element that is more commonly hidden. Hangers, lights, clothing and shelves could then be added to the frame to reconfigure the shop's space.

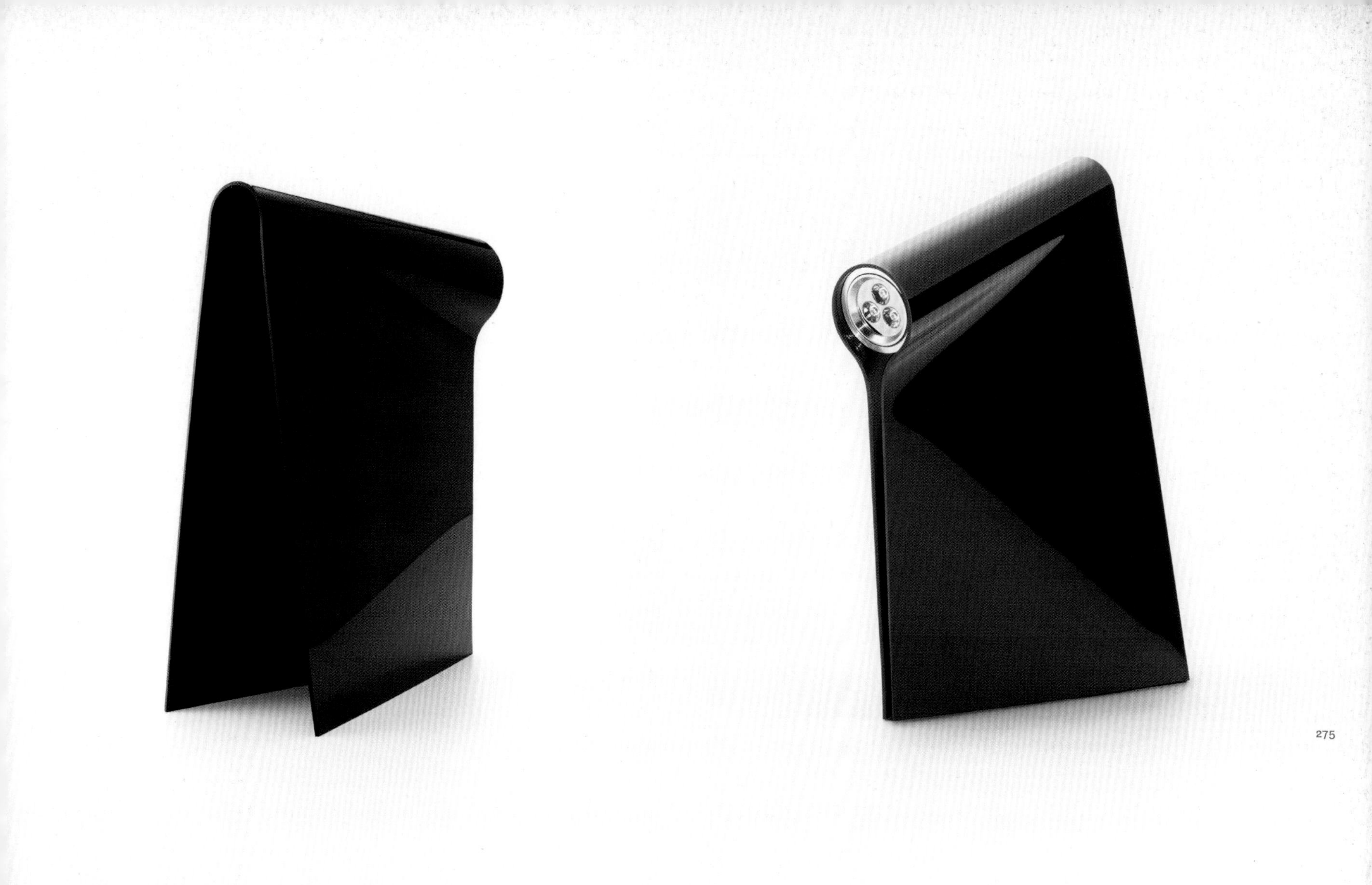

Hood

Established & Sons

A desk lamp whose light source is covered like a hood. A single sheet of acrylic serves as both shade and stand – casting bright light forwards and soft, indirect light to the rear. By moulding the acrylic tightly around the light source, it was possible to secure the fitting without the need for additional fasteners. The light fixture is the same as those used in spotlight fittings, allowing its angle to be adjusted by hand.

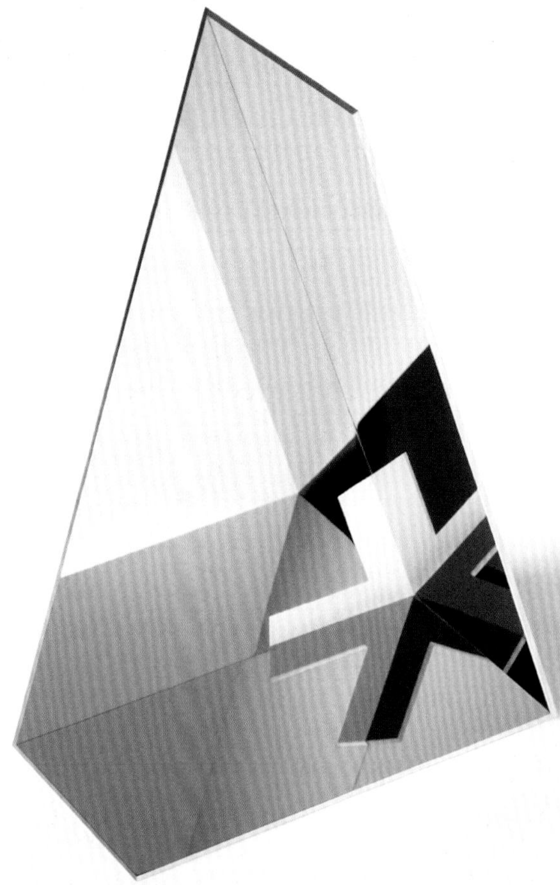

276

Trophy for Wallpaper*

Wallpaper*

A three-sided trophy that seems to have been constructed – like paper – from a flat sheet into a three-dimensional shape, with an angle of fold of 72 degrees, the same angle found between the spokes of the asterisk symbol in the *Wallpaper** magazine branding. Internal, mirrored surfaces – two of them facing each other at the crucial 72-degree angle – reflect a single, L-shaped, corner 'cut-out' in multiple directions in order to recreate the asterisk logo 'inside' the trophy. This kaleidoscopic effect creates beautiful patterns and colours, reflected from the surrounding environment, allowing the winners the enjoyment of deciding where to place their trophy.

Hitogaki
Elle Decor

An exhibition design for nominees of the Elle Decor Japan International Design Award. Rather than separating the space using walls, sixty cut-out panels of stylized human silhouettes created 'walls of people' that subtly demarcated each exhibit. Places where 'people' were gathered became display areas; those without 'people' naturally appeared as passageways. These human-like enclosures obscured the works on display from a distance, inviting visitors to come in closer and view them over each other's 'shoulders' like one might experience a street performance.

Starbucks Espresso Journey

Starbucks

A pop-up shop dedicated to Starbucks' espresso drinks. The goal was to create a link between books and coffee – both important parts of everyday life.

The space is designed to look like a library, with shelves seemingly holding books with covers in nine different colours. These are actually merely sleeves containing information about the various espresso types to which each colour corresponds. Visitors can stroll around the curving, free-form space, pulling 'books' off the shelves to 'read' and choose the drink that best suits them. At the counter, they trade the 'book' for an actual espresso drink but retain the cover, which they can use to customize their own Starbucks takeaway flask. The pop-out shape of a tall- or short-size tumbler insert has been punched into the reverse side of the book cover, which can then be used in a Starbucks Create-Your-Own Tumbler. The 'library' invites visitors to choose an espresso drink as they would a book, and verse themselves in the world of such beverages as though quietly journeying into a fictional world.

STARBUCKS ESPRESSO JOURNEY
STARBUCKS
ESPRESSO
JOURNEY

Bottleware

Coca-Cola

A tableware collection made from Coca-Cola's 'contour bottle' – a brand icon since 1916. Featuring its distinctive green tint, known as 'Georgia Green', and the fine air bubbles and distortions that are a hallmark of recycled glass, the collection reused bottles that had deteriorated over the course of extensive recycling and were no longer in circulation. Bowls and dishes were created that retained their distinctive lower shape, as though their tops had simply been sliced off. The base of each piece also preserved the original ring-shaped dimples that are particular to glass bottles.

Roopuppet

Rootote

The tote-bag brand Rootote is distinguished by the pocket, like the pouch of a kangaroo, located on its side. Many have different surface designs, but the side pocket – the brand's most distinctive feature – is usually the same. Here, the idea was to add an element of fun by turning something useful into something playful when not actually *in use*. We are very familiar with the feeling of reaching into a bag or pocket to retrieve something, but with the Roopuppet, instead the user's arm emerges from the inside of the tote, making this a very unusual specimen indeed. The Roopuppet comes in four versions: a kangaroo, a bear, a human being and a dinosaur.

Bamboo-Steel

Han Gallery, Taipei

A project commissioned by the National Taiwan Craft Research and Development Institute. Their brief was to discover sources for new designs through research into Taiwan's traditional bamboo handicrafts and furniture. The resulting chair, armchair and table set applies traditional methods of bamboo construction – such as weaving together thinly sliced sections and joining parts by wrapping them around each other – to tubular-steel pipes, which can be easily mass-produced with standard levels of quality. Such techniques also gave the hard metal a sense of pliancy.

Bamboo artisans visited the metal workshop regularly, and had many conversations about the project and its capacities. The Bamboo-Steel chair is the fruit of these conversations: a symbol of new futures for traditional crafts through the cross-fertilization of different techniques and materials.

Another technique was chosen for the Bamboo-Steel table design, in which lengths of steel pipe are sliced vertically into thinner sections, which are then woven together to create a surface – combining the legs and tabletop into one piece.

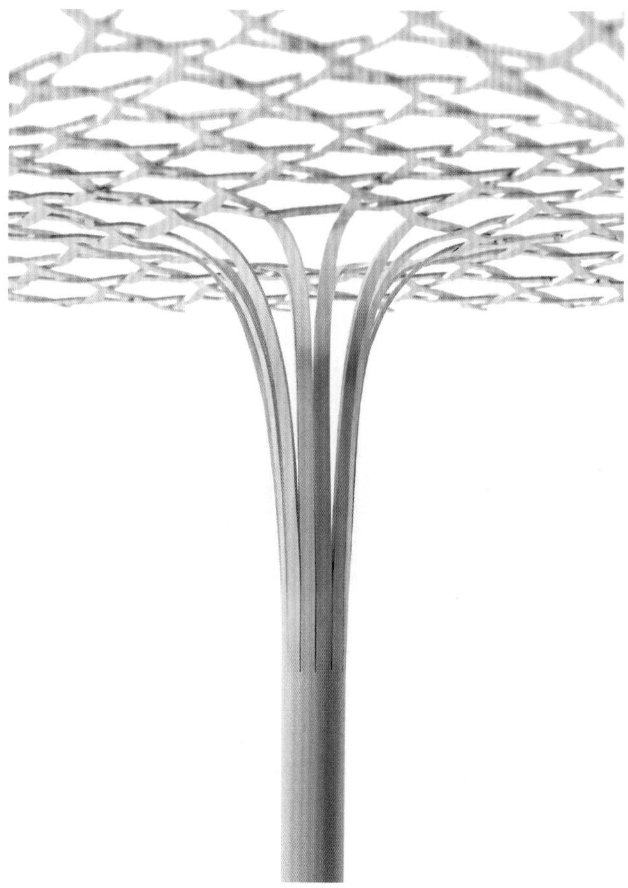

286

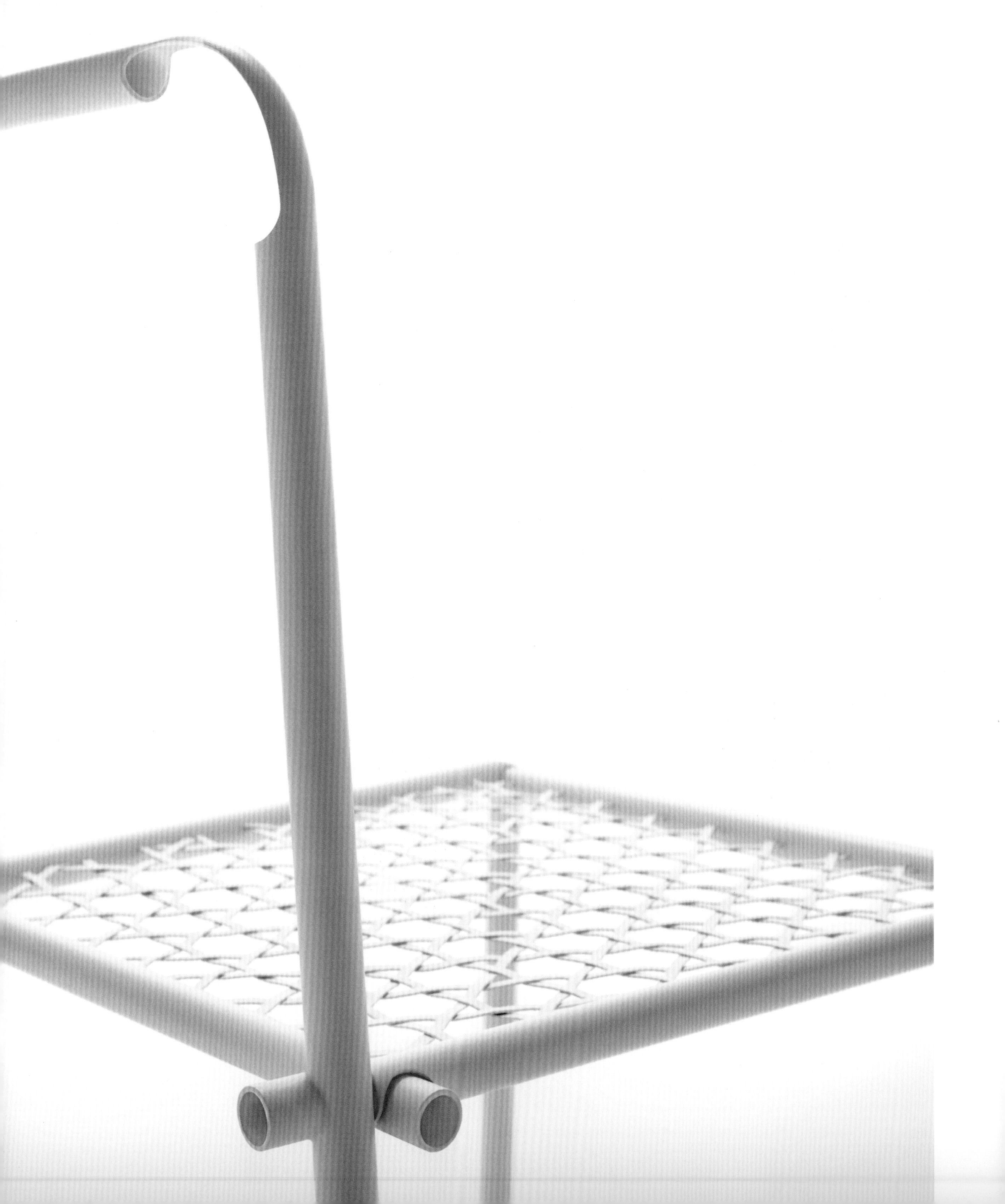

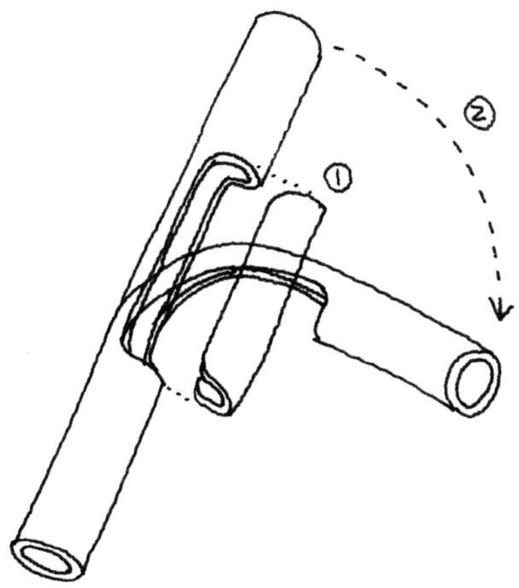
①
②

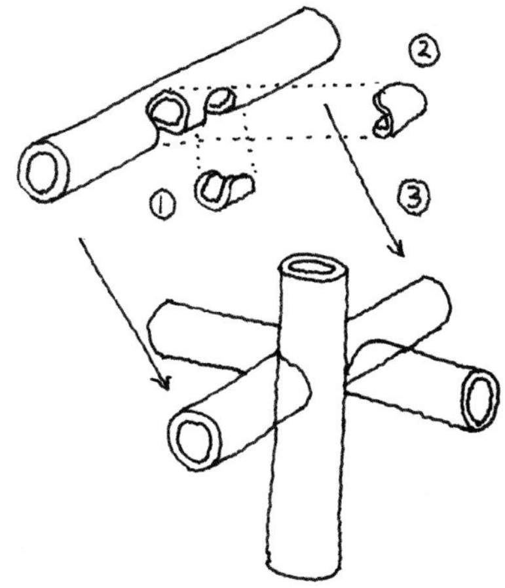
②
①
③

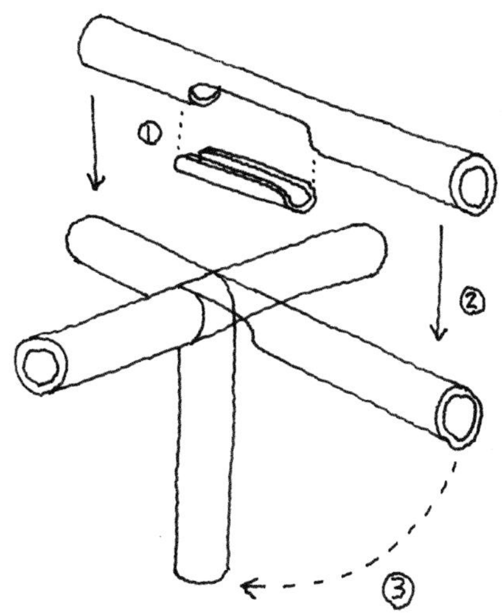
①
②
③

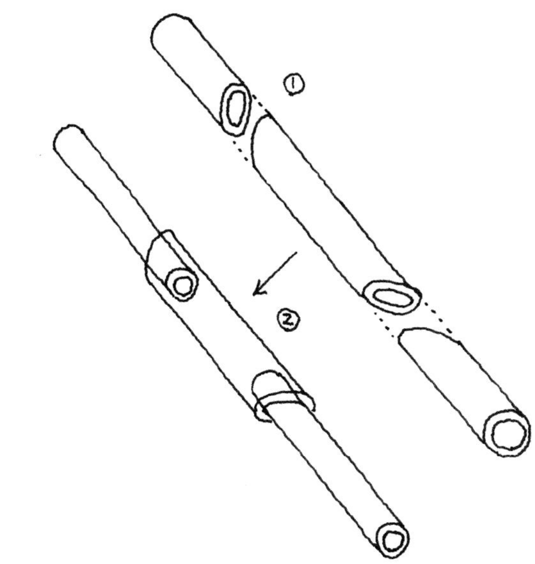
①
②

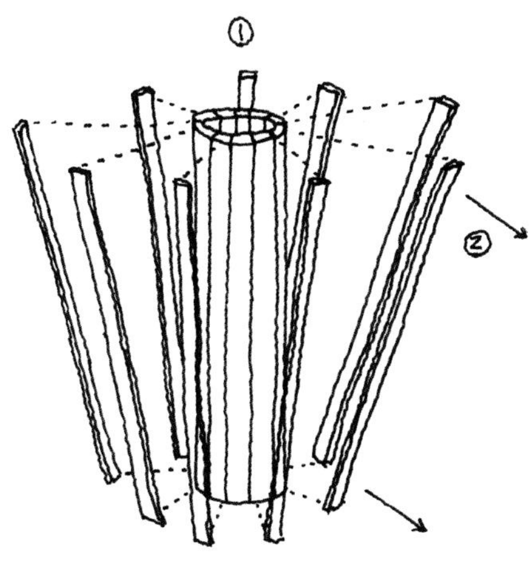
①
②

Press Lamp

Lasvit

A lampshade formed from a glass tube that has been gently pressed inwards at one end to create a pinch-like dent. This seemingly pliable shape eliminated the need for an internal support; its light fixture is wedged in the narrow gap created by the compression.

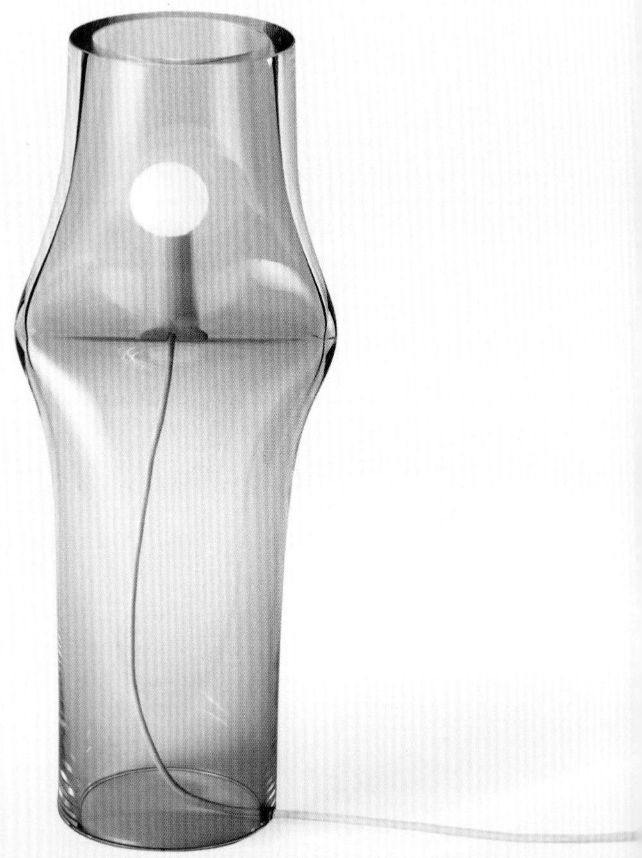

Inhale Lamp

Lasvit

Ordinarily, glass is made by blowing air into the molten material. The Inhale Lamp, on the other hand, was formed by blowing glass in the form of an air bubble and then sucking the air away again. Its design derives from these twin acts, its final form created by negative air pressure.

Bisazza Bagno
Bisazza

A bathroom collection comprising various elements of a bathroom suite, designed as though they were 'all in the bath together'. The feeling of connection that comes from bathing with strangers, at a hot spring or other public facility, is an important part of Japanese culture. Bisazza Bagno expresses this feeling through its design.

Taps are located within the 'curtilages' of the washbasin and bath-tub, partially submerged in water alongside accessories like soap dispensers and water glasses. A tray, also placed inside the basin, becomes an 'island' for objects needing to be kept dry. The washbasin stand encloses the bowl inside an open wood frame, through which the water pipe visibly rises up from the floor before turning into the tap. The mirror glass has been removed in places to mimic the effect of pools of water. This emphasizes the impression that the mirror, too, resides in an open, timber frame.

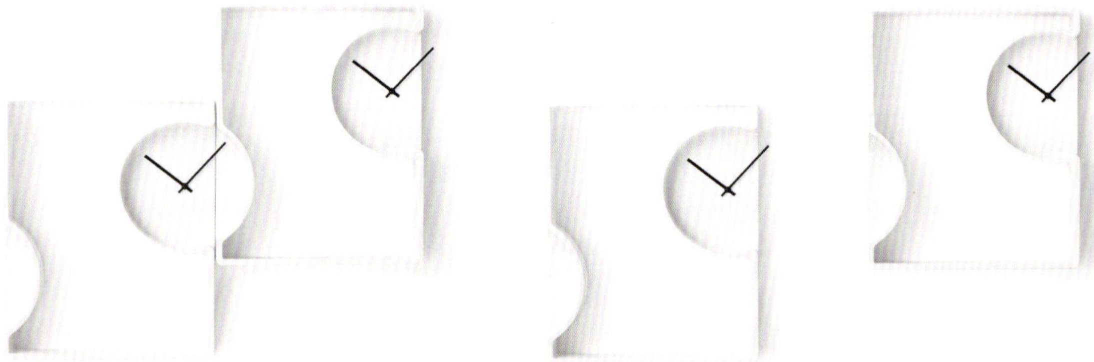

294

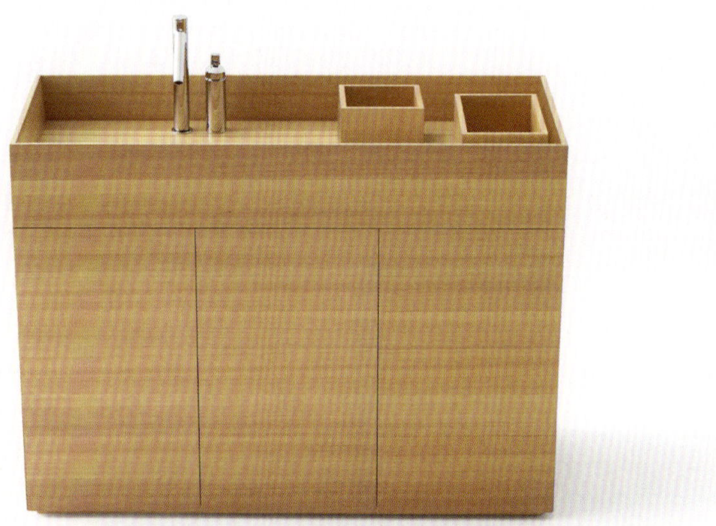

295

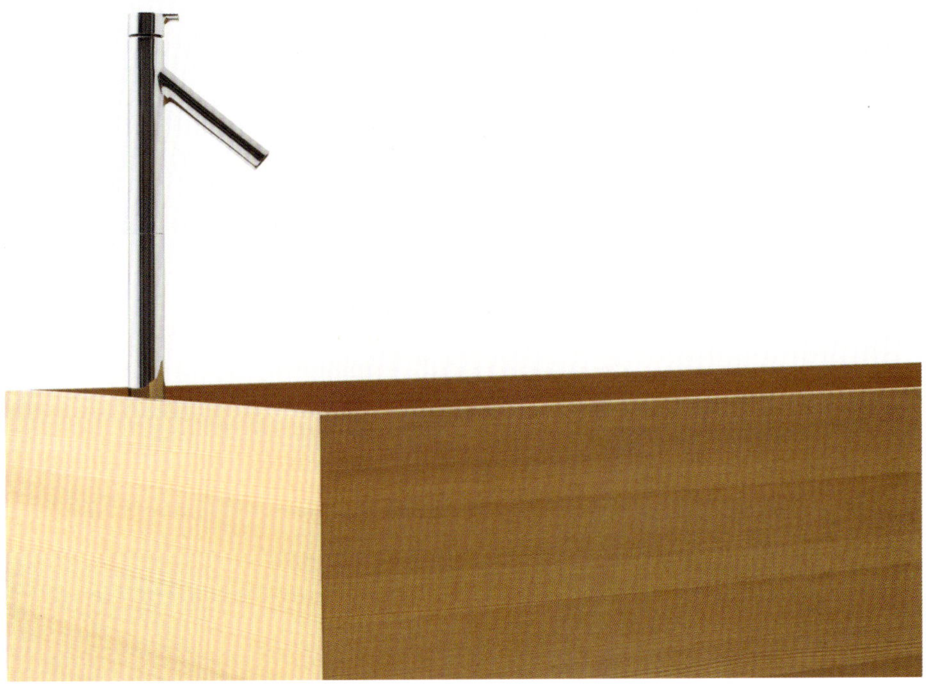

Mimicry Chairs

Victoria and Albert Museum, London

An installation for London's Victoria and Albert Museum, presented during 2012's London Design Festival. A semi-transparent chair with a seemingly soft, cushion-like back rest was pressed from punched metal, and placed in the main entrance and ten other locations within the museum including galleries, staircases and corridors. For each location, the chair was modified to reflect the particular space and objects alongside which it was placed.

On flights of stairs, rear legs were cut back to allow the chairs to 'sit' on the steps; elsewhere, legs were shared to allow them to 'huddle' together in groups – all to accord with their immediate surroundings. It was hoped that visitors might discover a new side to the museum by sitting on different chairs in unusual locations in order to observe, absorb and appreciate the objects in its collection.

Architecture

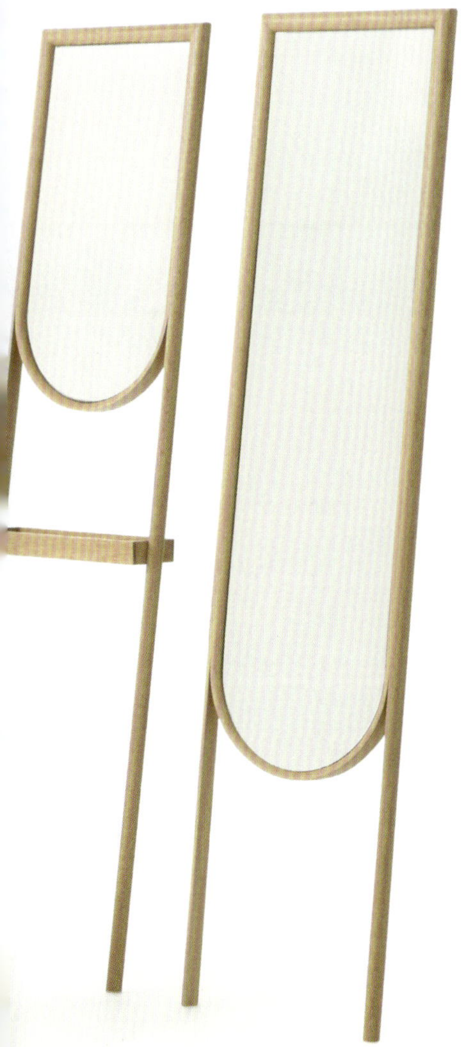

Splinter

Conde House

A furniture collection designed for a manufacturer based in Japan's famous Asahikawa wooden-furniture region. Each piece of timber in the set 'splinters', as though it is gradually peeling away. Chairs' backrests divide to become armrests and legs, and the top of the coat stand pares away to provide 'hooks'. The side table's stand splits at the base to form three legs. The larger pieces of wood retain their original thickness in order to provide strength where necessary, and thin pieces of wood that have splintered off are used for more delicate parts. During production, the timber was handled gently, going with the grain so that the material would maintain its original pliancy.

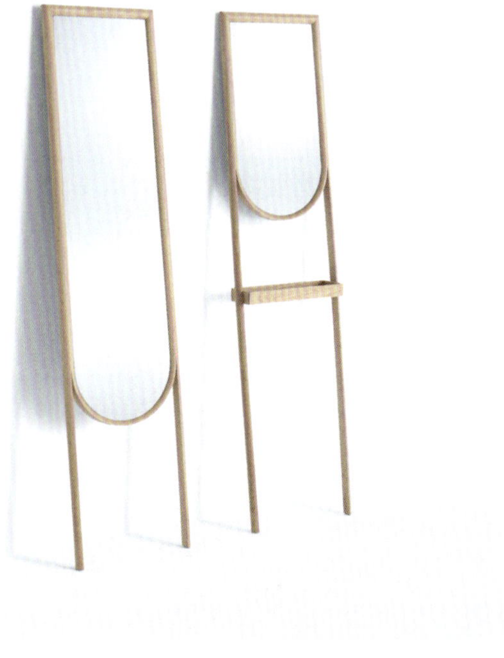

Aram

Gandia Blasco

Based on traditional Indian metalwork techniques, the Aram collection of stools and low tables comprise a metal frame, around which thinner steel wires are woven. The spacing between the wires widens gradually from top to base, creating a gradation effect.

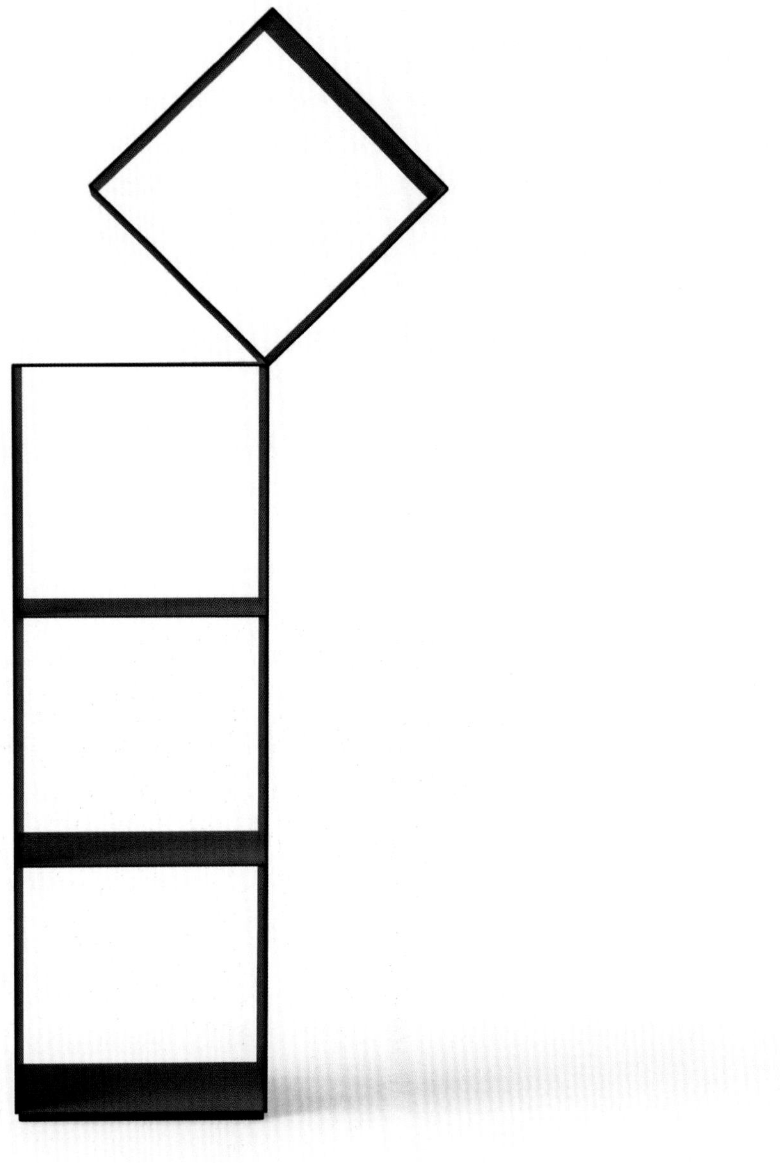

Drop
Cappellini

A bookshelf whose uppermost section is tilted at an angle to provide a variety of options for storing books – they can be draped over the angled top or stacked on a slant, for example. The backs of the shelves were removed so that Drop can be used from either side, allowing greater flexibility when positioning it in a room. The bookshelf was designed in three sizes: comprising two, four and five shelves.

Shivering-Bowls

Triennale Design Museum, Milan

A collection designed for the *KAMA. Sex & Design* exhibition at the Triennale Design Museum in Milan. The curators asked eight designers to create an object, in conjunction with the exhibition that explored ideas of Eros in design from ancient times to the present, from a cultural anthropology and mythical perspective.

In response, a series of extremely thin silicone bowls was developed that focused on the physical and spiritual pleasure provided by the feel of an object. They resembled ceramic bowls but provided a tension to accompany this perception, which was generated by an exceptional thinness that would have been impossible to achieve with clay. The bowls change shape as easily as liquid when they are touched, and continue to quiver momentarily in response to an outside force. They were intended to express Eros through a design that invoked desire – one that viewers simply couldn't bear not to touch.

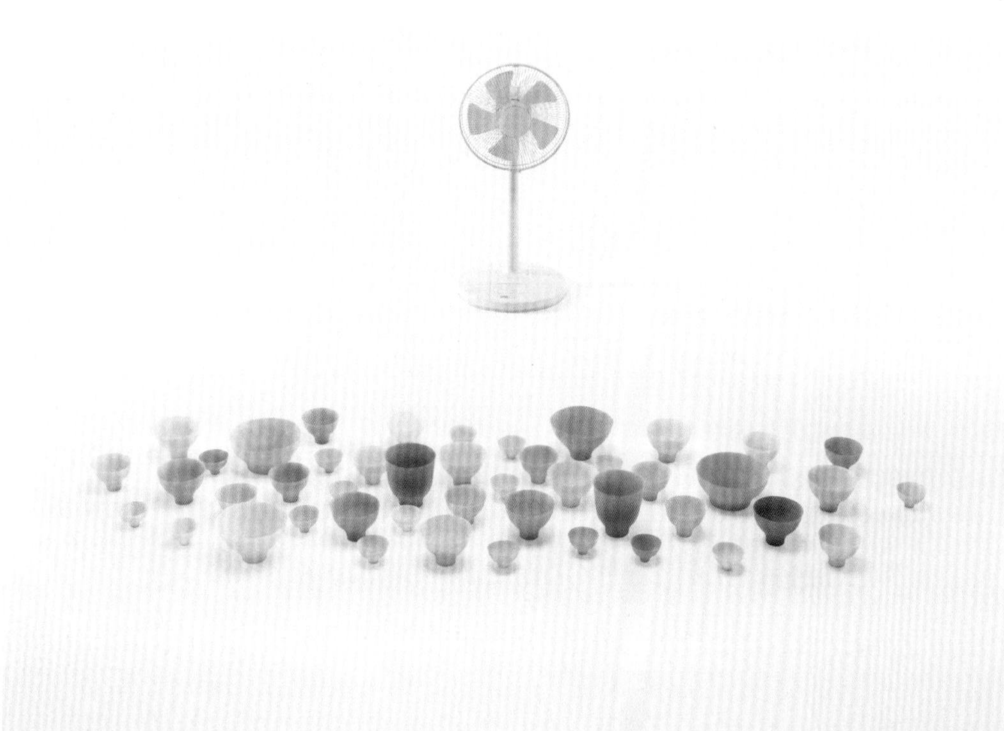

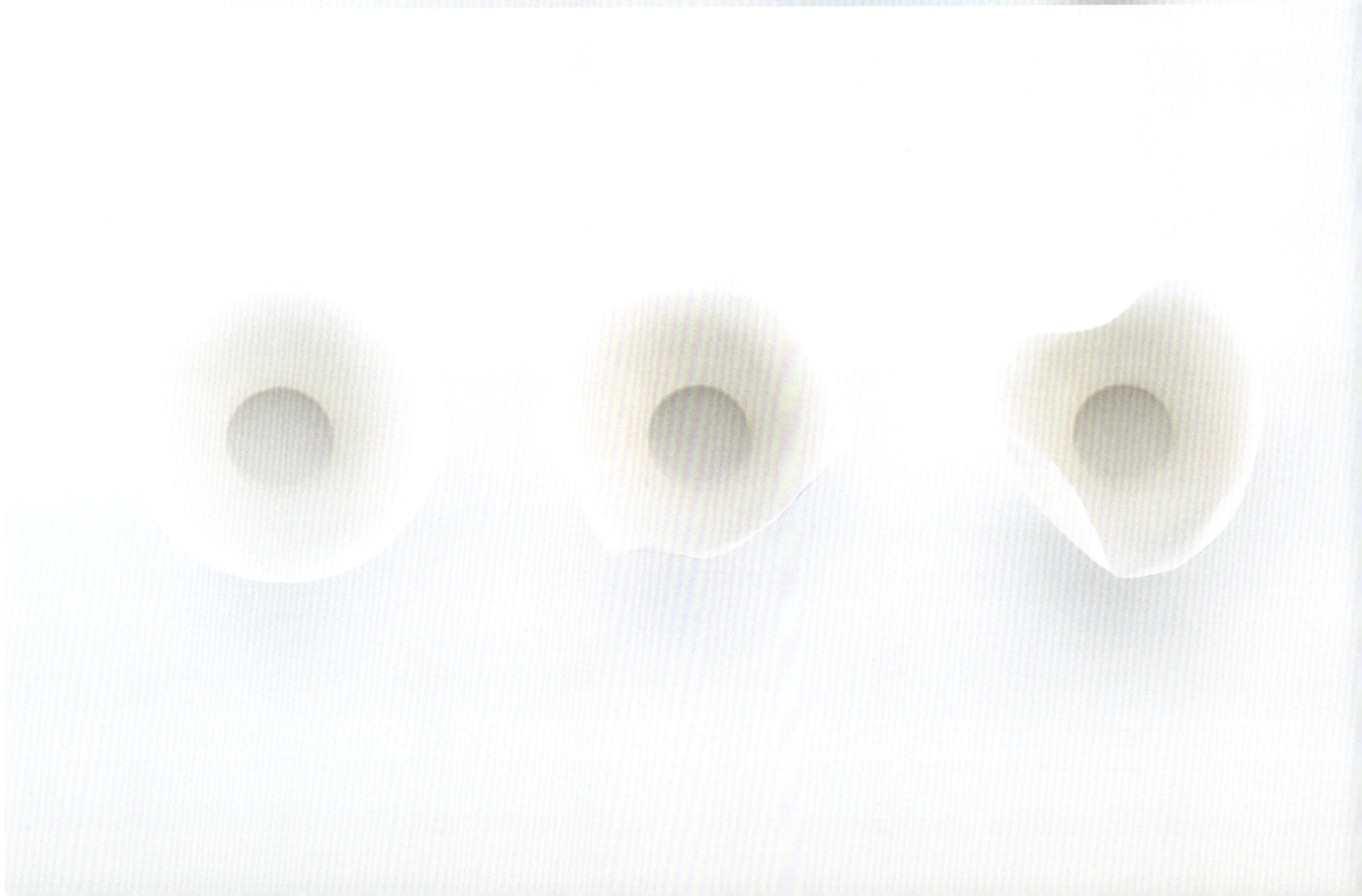

Fishline Chair

Biennale Interieur, Belgium

The Fishline Chair is an exploration into methods of finishing wood surfaces that extend beyond the usual applications. For the design, fishing line was wound tightly around all the surfaces of a wooden chair, to give it a new lustre and a fine unevenness. The line was dyed before use, allowing it to function as a kind of varnish, bringing the wood grain to life while adding colour.

Camper 1st Concept

Camper

The *raison d'être* of Camper shoes is neither to help people run more quickly nor to give the wearer recognizable status through ostentatious branding, but simply to help people to enjoy walking. Accordingly, this interior shop design displays the shoes across its walls and ceilings as though they are floating in the air and strolling freely around the premises on their own. This effect was achieved, in part, through the installation of steel walls and fitting the shoes with magnets so they could be fixed easily in place.

Shoes are also displayed on floor-, ceiling- and wall-mounted plinths, where they seem to float on 'footprints' supported by thin pipes that rise from the surfaces.

CAMPER

Bird-Apartment

Ando Foundation

A treehouse designed for the Momofuku Ando Center, a facility devoted to promoting and increasing access to outdoor activities. The centre is located in the forest of Komoro City, in Japan's mountainous Nagano Prefecture. The Bird-Apartment treehouse provides 'collective housing' for many birds and for a single person. On one side, it has entrances to seventy-eight nest spaces for birds. The other side has an entrance, accessed by ladder, for one person, who can look into the nests through peepholes inside the treehouse.

Catalogue of Works

2009

Shortcut
Boffi

A towel hanger made by bending one metal rod in six places and fusing it in two. The aim was to create a functional object from the minimum number of materials and processes.

Shupatto!
ST Corporation

A redesign project for a range of electronic deodorizing devices, which are fitted with a spray can that automatically releases its fragrance at regular intervals. To avoid the garish and gadget-like aesthetic normally sported by such appliances, the redesign adopted the familiar shape of a bar of soap, instead evoking a feeling of cleanliness and approachability. By removing unnecessary parts, and combining the release button and operating light into one feature, the design was able to reduce the size of the units by 25 per cent, as well as minimize production costs.

Collar
Quodes

A small storage unit comprising 250 x 200-millimetre (10 x 8-inch) sheet-metal modules that can be freely added to and rearranged. The edges of the metal surfaces extend outwards, similar to a raised shirt collar, serving as door and drawer handles. They also give the units a particular ambiguity, as though they have no up or down, left or right. The cabinets, drawers and shelves work together naturally, giving the collection a distinct sense of unity.

Asobio 2nd Concept
Cobest

A second retail-space design for fashion brand Asobio. The design continued the 'photo studio' theme established in the first store by incorporating monochrome images onto the floor and walls. These varied in size to imitate the effect of a camera's zoom lens, and in sharpness to recall the sense of being brought into and out of focus. These effects enhanced the space with depth and helped the displays to stand out by appearing 'in focus' in contrast to the shop's often blurred decorative patterns.

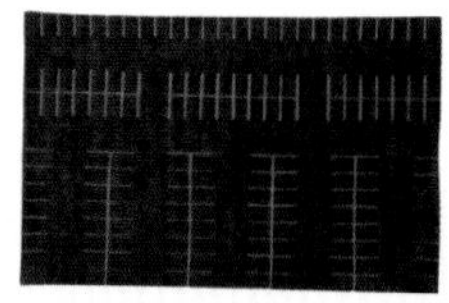

Parking
Arketipo

A patterned carpet that looks like an aerial view of a car park. The idea is that one could 'park' oneself, a magazine rack or a side table on the carpet for a short rest, transforming it temporarily into a relaxing space outside the busy flow of the everyday.

Kub
Moroso

With their varying dimensions, and digitally produced wood-bark patterns, the pieces in this furniture series recall *kub* (tree stumps). The solid, linear, metal pieces represent a visual play on 3D-software 'wireframes', in which the many vertices of a designed object are visible as though it had no 'skin'. The soft, smooth, fabric-covered versions project a strong sense of solidity, but on closer inspection the pattern is visible on their surfaces. By combining these oppositional elements – line and surface, soft and solid – it was hoped to generate objects that exist in an ambiguous space between natural and digital expression.

Dew
Moroso

A round pouffe with the plump fullness of a droplet of dew. Strips of leather of varying widths were woven together to add an unusual three-dimensionality to the pouffe's surface. Reflecting the way in which an image becomes distorted in a water droplet, Dew's latticed pattern shifts across its stretched leather surface.

Hexagon
Quodes

A bookshelf designed in the shape of a hexagonal pillar. Each open compartment is set off 120 degrees from its neighbour, allowing access from all directions and giving Hexagon the flexibility to be placed in any position in a room. With two different heights available, the bookshelf can also be used as a side table, a pedestal for displaying other objects, or even in multiples to create a freestanding room divider.

2010

Fit
Takata Lemnos

A flower pot that is recessed at both ends to allow plants to be added from either side. The cylindrical design gradually widens at one end so that when two or more are laid out in a line, in alternating fashion – one 'end' up followed by the same 'end' down, and so on – the pots interlock perfectly in section.

24 Issey Miyake 2nd Concept
Issey Miyake

A retail display featuring Issey Miyake's Baobao bag, which has no set form but settles depending on how it is placed. The design abandoned the hard, smooth fixtures found in most shops, and introduced variable-height alternatives made of thin, steel rods that stand like fields of metal grass in the store, adopting random arrangements like the bag. Supported thus by 'points' rather than surfaces, the bags seem to waft in the air, changing shape and expression according to the uneven terrain beneath.

Yokai Daruma
Idea International

With its Japanese-folktale spirit-monsters, or *yokai*, the manga and anime series 'GeGeGe no Kitaro' has been immensely popular since launching in 1968. This series updates the traditional Japanese Daruma doll – most often used as a good luck charm – with 120 new designs based on *yokai*, including many characters from the series. Since ancient times, Japan has been rich in stories about *yokai* as manifestations of natural phenomena and chance events. Equally laden with symbolism, the Daruma could be said to share a close affinity with the *yokai* – thus providing the perfect opportunity to fuse two icons of Japanese tradition.

Virus Attacker
ST Corporation

A diffuser that ionizes water and oxygen molecules through plasma discharge (the release of an ionized gas by passing electric current through the gas), and uses the resulting oxygenation to disinfect, deodorize and eliminate viruses.

Its inbuilt fan aims concentrated ion streams at doors, windows and hallways, where viruses and bacteria might enter. The cylinder's neck is angled forwards, enabling it to pinpoint strategic areas. This angling also makes the top the most accessible surface, so the indicator light and whole-face on/off button are located there.

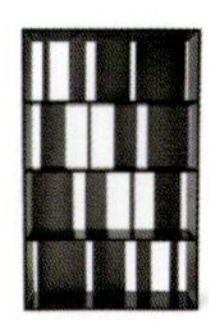

Sundial
Kartell

The appearance of objects and their shadows are slowly transformed as the sun moves across the sky, casting its light from different angles over the course of a day. The shelves of the Sundial bookcase are all arrayed at slightly different angles, like the shadows cast by this ancient time-telling device. As a result, the appearance of the books placed on them, and the look of the bookshelf itself, are also always changing.

Crystal-Puzzle
Swarovski

Swarovski's crystal components are created by repeatedly making linear cuts into a large piece of crystal. The Crystal-Puzzle allows people to experience this unparalleled precision by taking apart blocks of crystal from a larger cubic construction, allowing the sparkling gem inside to emerge. Reassembling the cube is equally enjoyable, much like putting together the pieces of a puzzle. This process of discovery and reconstruction encourages a deeper appreciation of the beauty of Swarovski crystals.

Three
Acro

A packaging design for Japanese organic cosmetics brand Three, whose name refers to the keywords describing its fundamental values: natural, honest and creative. The motif for integrating these core values into the packaging was that of blocks carved from stone – a natural material shaped patiently by human hand. The different bottles can thus be stacked horizontally or vertically, and fit together perfectly. The rubber finish is responsive to the touch, and the edges are slightly rounded as though worn away over time. Two different, warm shades of grey were used to distinguish between the skin care and make-up lines.

Dent
Arketipo

A low, table with a literal, cubic 'dent' in one side. The recessed surface area provides a natural resting place for objects like books, flower vases or lighting fixtures, almost like a 'trouser pocket' for a table.

Clear Perfume Bottle
1% Products

A perfume bottle filled with nothing. When part of the cap is pulled up, an atomizer capable of holding 2 millilitres (0.07 fluid ounces) of perfume appears. Though the acrylic container contains no perfume, it is nevertheless a solid object and can therefore be perceived as being both empty and full at the same time. Bubbles inside the bottle add to the illusion of a liquid content. Through this experience of 'emptiness', users are invited to explore the invisible sensation that is fragrance.

Fatback
Tectona

An outdoor lounger with an unusually large backrest. It is solid and comforting – producing an effect like leaning up against a rock or a massive tree – but is also hollow. The integrated lounge chair and storage design adds functionality, allowing cushions and other items to be stowed when not in use.

Ceramic-Speaker
Take Action Foundation

A collaboration with Mitsuke Masayasu, a potter working in the tradition of red-glazed Kutani ware. Mitsuke's elaborate lines were fused with digital-manufacturing techniques, creating a set of high-end audio speakers in wafer-thin ceramic. The computer-controlled manufacture of this substance shaves thin slices from ceramic slabs, fixes them with mercury vapour and mounts components with a robot arm; human hands play no part. It was hoped that Mitsuke's patterns would 'disrupt' the process – allowing the surface to acquire a new, decorative role.

2011

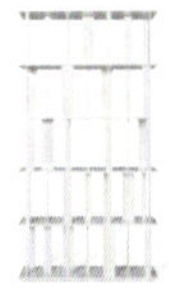

Forest
Driade

A bookshelf comprising a 'forest' of supports of different widths and depths. All the supports have a standard cross-sectional area of 9 square centimetres (1½ square inches) – so the wider ones are shallower in depth, and vice versa. As a result, the bookcase also resembles a forest of trees at different stages of growth, in which books and objects can 'wander' like animals between them, seeking shelter in different places.

Orime
Elecom

A wireless, laser mouse constructed entirely from flat surfaces, rather than the usual multiple curves. Formed as though it were a piece of folded paper, the *orime* (fold) mouse fits comfortably into the hand. It also resembles low-resolution CG data for computerized pixel shading – as if sprung out from the screen for a design that seems neither digital nor analogue, but possibly both or somewhere in between.

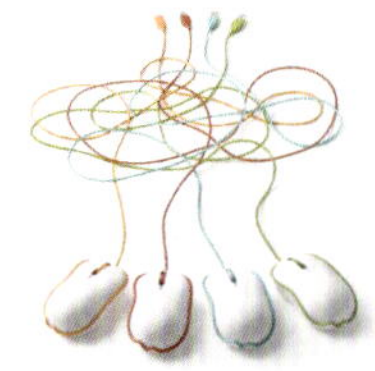

Rinkak
Elecom

In Japanese, *rinkak* refers to an object's outline – as seen in a quickly sketched, freehand drawing. In this design, the brightly coloured mouse cable that snakes out from a computer reflects the outline of the mouse itself, becoming its *rinkak*. The mouse, in effect, becomes the background, and the cable – until now a supporting actor – takes to the main stage.

Otokurage
Elecom

A design that enlarges the silicone caps that cover earphones in order to hide the entire device. The resulting set of in-ear headphones has a soft feel and unusual transparency, reminiscent of a *kurage* (jellyfish). The name of this sea creature was paired with the Japanese *oto* (sound) to create the title for the product. Because its components are unified, substituting the colour or size of the ear caps changes their look entirely.

Catalog
Swire

Catalog is a Hong Kong-based sports-fashion retailer and the Beijing store was its first foray into China. The company's name reflects its philosophy: treating different brands equally, and actively suggesting ways of coordinating items from various lines. The store design sought to recreate the specificity and attraction of this retail experience by creating a series of window displays that were composed like pages from a fashion magazine. Rather than simply lining the facade of space, the framed displays continued throughout its interior, creating a spatial experience similar to flicking through a catalogue.

Bell-Orgel
Isetan

A series of wooden bells containing musical-box mechanisms, designed for 'Japan Senses', an advertising campaign led by Tokyo's Isetan department store. In cooperation with composer Sakamoto Ryuichi's More Trees initiative – which promotes the protection of forests as absorbers of CO_2 – the project commissioned fifty-seven designers to customize the bells, which were made from Japanese *hinoki* (cypress) harvested as part of sustainable forest-management practices.

The bells were designed in three varieties: a hand bell, designed to stop playing its melody when laid on its side; a standing bell, which, on turning, winds the musical-box mechanism, making it play; and a hanging bell.

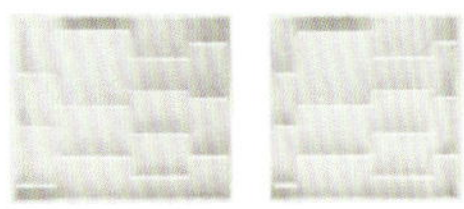

Shift
Lema

A set of bookshelves that divide and scatter at different heights on a vertical axis, creating many small spaces of differing proportions. The surrounding frame isolates these spaces from the exterior world, emphasizing their particular spatial sensibility like a *hakoniwa* (miniature garden).

Japan: Tradition. Innovation.
Canadian Museum of History, Quebec

An exhibition design for *Japan: Tradition. Innovation.*, held at the Canadian Museum of History (formerly the Canadian Museum of Civilization), Quebec. Using a comparison between contemporary and Edo-Period (1603–1868) Japan, the exhibition aimed to explore the relationship between Japanese culture and design.

In the massive 970-square-metre (10,440-square-foot) space, Edo-style 'roofs' of varying sizes and heights, rather than interior walls, demarcated the exhibits. Areas under larger, lower roofs formed 'closed' spaces, from which visitors looked out onto 'outside spaces' and – conversely – peered beyond into other 'inside spaces'. This resulted in a hybrid, changing space in which the concepts 'interior' and 'exterior' proved elusive – an experience similar to that of Japanese urban space.

Onb
Yoshida & Co

A collection made in collaboration with Yoshida Kaban's Porter brand. Onb comprises five items: an A3-sized portfolio case, A4 and A5 file/computer cases, an A5 organizer cover and a pencil case. The portfolio case folds down to A4, and a thin board can be added to its outside pocket to increase rigidity. An internal pocket in the file cases accommodates foam padding or thin board, allowing their customization. The pencil case easily attaches to the inside or outside of the larger cases. The collection's name references the word *onbu* (piggyback).

Halsuit
Haruyama

The great strength of Haruyama, a high-volume men's-suit retailer, is choice. However, the number of options can often overwhelm customers, and the atmosphere isn't always as conducive to relaxed shopping as the company would like.

This design moved the fitting rooms from the edge to the centre of the shop floor, using their exteriors to showcase coordinating options. Most suits are for the office, so the lighting is reminiscent of desk lamps, and office-style shelving creates the right atmosphere. Shoppers purchase at a 'reception desk', and lounge and conference room-type spaces help them to visualize their work styles as they select their suit.

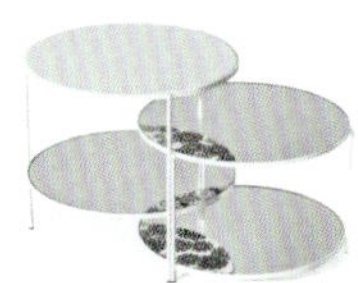

Pond
Moroso

A set of low, two-tiered tables with mirrored surfaces and undersides printed with a graphic pattern. Available in four sizes of varying heights, the tables can be placed together to create an interplay of surface levels and patterns, as one table's mirrored surface reflects another's underside. The product was named

Pond after the way in which it reflects its environment.

Indulgi
Moda En Casa

An interior design project to transform a small clothing shop in Kyoto's Nakagyo district. Originally occupying a deep, open space, the shop was subsequently divided in order to interrupt its sightlines and create a level of mystery and intrigue for the customers.

By adding various shielding elements, the interior can never be seen in its entirety – different aspects appear and disappear from view, changing customers' experiences as they move about. In an additional twist on perspective, more doors were installed in the space to further disrupt the flow.

2012

Zabuton
Moroso

A lounge chair that takes the form of a Japanese futon mattress casually draped over a metal frame. Its seemingly random folds are abstracted and cast in polyurethane foam, creating a distinctive and gentle pattern of 'controlled creases'. This material allows the chair to maintain its slouched form while also being soft and comfortable to sit on. The lounge chair comes in three versions – featuring one arm, two arms and no arms – and can be combined to create a longer/larger seating unit.

Cloud
Moroso

A set of tables perforated with a halftone pattern generated from photographs of clouds. After abstracting their shadows into a series of dots, the cloud imagery was applied to colour-coated steel tabletops using laser hole-cutting technology. The three tables can be used very flexibly: smaller ones can be placed on the largest one, tucked away underneath it or brought out completely for separate use.

Wandering Forest
Hermès

Window displays designed for the Hermès boutiques in Yokohama and Tokyo. Created from a combination of two dimensional tree-shaped screens and three dimensional table units in a variety of colours and heights, the displays playfully experimented with depth and perspective. The furniture was interspersed between the screens so that some areas were visible, while others disappeared into the surrounding 'forest', resulting in an ambiguous transition between foreground and background.

Zipper
Nava Design

A bag comprising individual components that can be zipped together. It is easily customizable, depending on which parts are used; modifiable elements include the bag's main body, an interior pocket and a notebook/PC case.

Dark Noon
Noon Copenhagen

Danish company Noon's wristwatches use an exclusive system in which turning discs rather than hands tell the time.

The discs and dial for Noon's limited-edition No. 17 watch adopt opaque, matte colours, black faces and white lines. Accordingly, the discs' movement changes the wearer's perception of time's flow. The collection features five designs: 'times' has twelve small inner faces; 'numbers', whose parts meet at 3, 6, 9 and 12pm; 'chrono' borrows classic chronograph elements; 'line' uses dotted lines for hands; and 'gear' has a cog-like appearance.

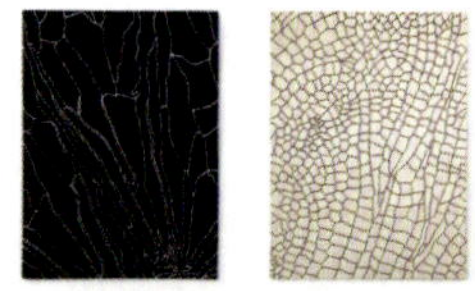

Butterfly/Dragonfly
Gandia Blasco

A rug with a pattern of magnified butterfly and dragonfly wings. Rather than artificially creating a new design, the scale of an existing, natural form was modified, providing a new way of looking at the natural world.

Rinascente
Rinascente

A renovation project for the womenswear floor in Milan's iconic Rinascente department store. To give visitors a sense of the city, seventeen large, traditional Milanese windows were installed as space dividers. Inspired by encounters with shop fronts, courtyards and public squares, a film was applied to these windows rendering them opaque when seen from an angle but transparent when viewed directly. This allowed clear visibility of the products when walking through the aisles but also a level of privacy when stood in front them. Lighting fixtures recalled urban examples, strung from cables between buildings, and in-store mini-boutiques were accessed through external-style, arched openings.

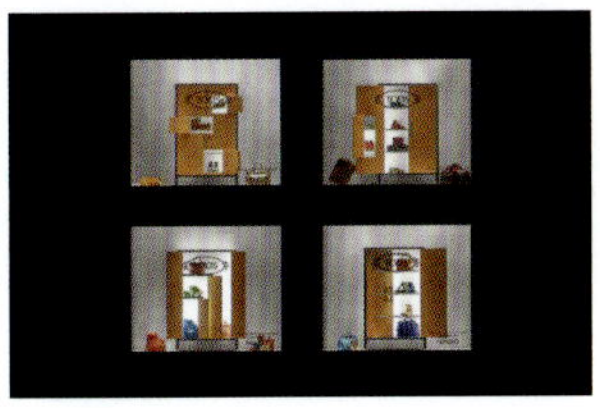

Cabinet in the Window
Tod's

Window displays for fashion brand Tod's, shown during 2012's Salone del Mobile event in Milan. Ultimately, the Salone is a furniture fair, so each window display was created as a single, large piece of furniture – a wardrobe – made from Tod's characteristic leather and mirrored stainless steel. The other common feature was a playfulness with doors and openings. The wardrobes opened in places that don't usually have openings, had doors that opened only to reveal further doors, or had doors that only opened inwards.

Pencil Chair
K%

A collection of matte-black painted oak furniture, comprising an armchair, a dining chair and a lounge chair. The legs of each piece taper at the end (and, in the case of combined leg and uprights, at both ends), like a pencil that has been whittled with a knife.

Warp
K%

The single curved surface that runs through the centre of these otherwise flat-planed, box-like tables and stools plays a structural role by stabilizing them. It also has the visual effect of pulling viewers into the furniture's shadowy interiors.

Harcourt Ice
Baccarat

French luxury brand Baccarat has been making crystal since 1764. This redesign of its signature Harcourt line strove to highlight a new facet of the material by creating a form like perfectly frozen ice, then 'melting' it. Baccarat's skilled craftspeople smoothed the edges of finished Harcourt glasses by dipping them into the acid ordinarily used in the final stages of the polishing process. The soft feel and light refraction of these melted edges captured the moment at which ice dissolves into free-flowing water – and represented the gently rounded edges of Harcourt glass after many years of use.

Aroma Cup
Häagen-Dazs

An aromatherapy candleholder made for Häagen-Dazs, who commissioned five designers to create objects based on five different ice cream flavours – in this case, vanilla. The design recreated the ubiquitous paper cup, lid and accompanying plastic spoon in ceramic – adding 'melting ice cream' details for another touch of realism. The candle inside gradually releases its essence as it burns, reminding users of the way that ice cream melts in the mouth, while releasing its delicious flavour. When unlit, the lid can be closed and the spoon used as an aromatic-oil diffuser. Aroma Cup takes the pleasure of tasting ice cream and resituates it within our senses of sight and smell.

Airleaf Mini
Mikuni

An air-purification device, whose form recalls a new shoot sprouting from a small pot, that attaches to the outlet vent of fans and air conditioners. It contains tiny beads impregnated with essential oils extracted from fir trees. When air flows through these beads, it reduces and removes carbon dioxide in the atmosphere and restricts the production of antioxidants. Air purification increases as the 'new shoot' grows, allowing users to adjust its size, and therefore its scent, intuitively, and providing a visual measure of air purification.

Cupnoodle Forms
Cup Noodles Museum, Yokohama

A novelty souvenir range created for the Cup Noodles Museum in Yokohama. The instantly recognizable form and branding of this ubiquitous Japanese convenience food was chopped, squashed and stretched to create a series of uncanny replicas. Even subtler changes to the original cup-noodle container give it a very different appearance, reminding us that the line between the ordinary and the extraordinary is extremely fine.

Mist
KME

A series of low tables designed for KME, a manufacturer of copper products. Rather than exposing the material entirely, all but one section of the tables was painted white, using a subtle gradation to partially reveal the copper undertone, and thus piquing the viewer's curiosity.

2013

80 Sheets of Mountains
Stockholm Furniture & Light Fair

An installation created for the entrance hall of the main exhibition space at the Stockholm Furniture & Light Fair in 2013, at which nendo were named Guest of Honour.

Eighty sheets of 5-millimetre (3/16-inch) -thick polystyrene were laser-cut, stretched and plugged into holes in the floor, forming a series of partitions. These were arranged to create a 'landscape' of snow-capped peaks in the space. Care was taken to minimize the exhibition's environmental impact. The recyclable polystyrene was modelled on location so that only one delivery truck was needed (for the thin, flat sheets), and the 'mountains' themselves could be easily stacked for clearing from the site – an operation that took roughly one hour.

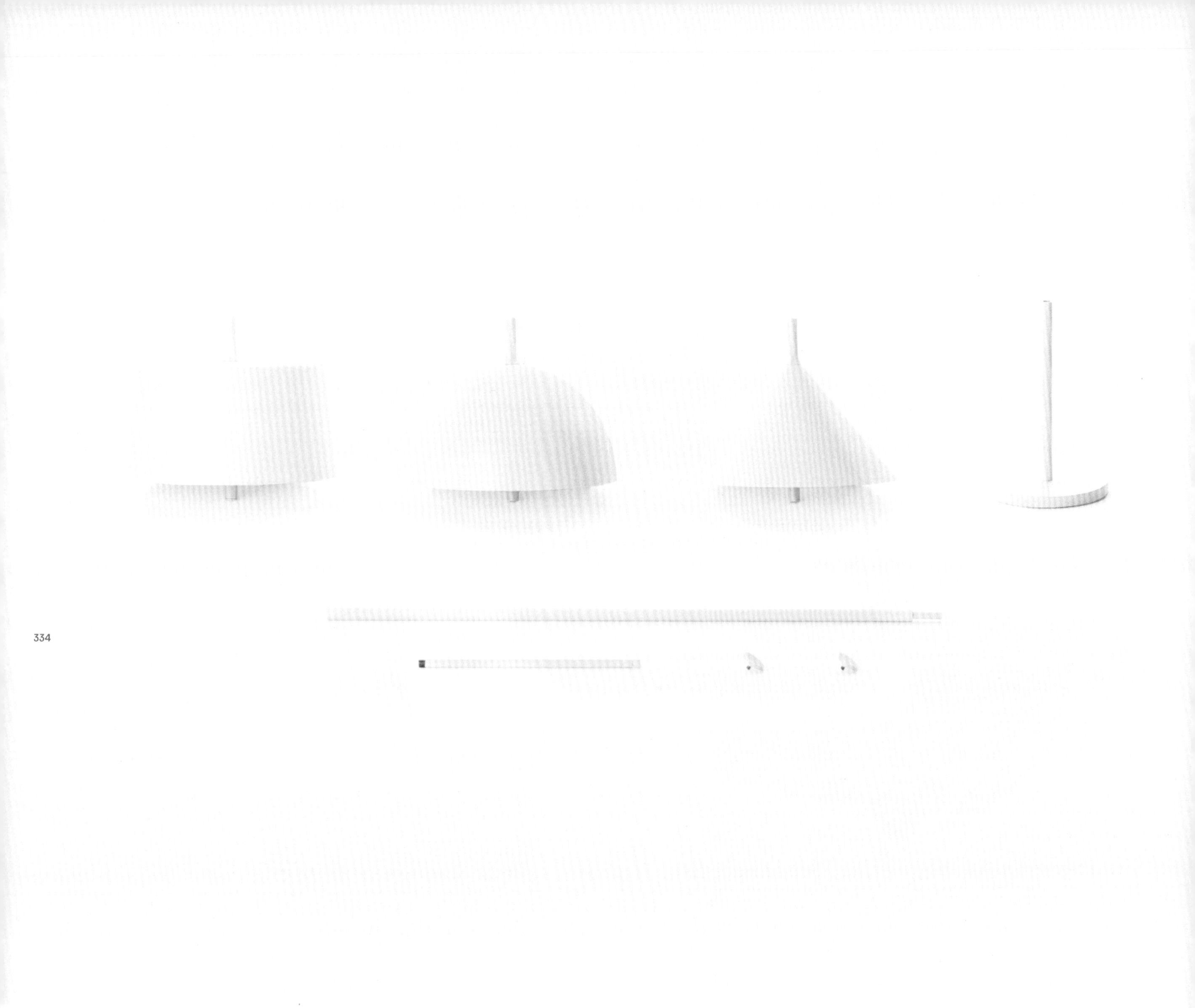

334

Nendo W132
Wästberg

A lighting fixture that is created in the process of assembling its constituent parts: the shade, post and stand. The height and form of the light can be changed easily by adding and re-arranging its components. By attaching the wire unit to the shade, it becomes a pendant lamp. Adding a longer pole to the desk lamp turns it into a floor-standing unit. The shade is available in three different shapes – a cone, a half-sphere and a cylinder – and the collection also encompasses units that become side tables and other accessories using the same poles and pedestals.

Fold

Conde House

A shelving unit that is made of perpendicular, interlocking wooden boards that, in alternating rows, create openings to the left and right. Its zig-zag formation means that books placed within the recesses face diagonally outwards in two directions. Thanks to its superior craftsmanship, the joints appear seamless – as though single boards have been bent and interwoven like paper chains or textiles. Looking at the Fold shelves, it is virtually impossible to know how its parts are connected.

Nuno

Vibia

A collection of resin-formed lampshades adorned with layers of *nuno* (cloth). The lighting fixtures blend in like decorative objects when turned off, and create complex shadows when lit. Some areas of the shades are covered with only one cloth layer, while others have up to twenty-four – providing a tonal range of up to two dozen gradations for the shadows cast. Rather than being arranged geometrically, the cloth layers are varied, creating a complex surface with a particular softness.

Stone Garden

Caesarstone

An installation featuring clusters of one-legged tables that use each other for stability by overlapping their surfaces. As no nails or screws were used to secure the tables to each other, the depth and direction of these overlaps could be changed as required, and the groupings in the room could be shaped as freely as an amoeba changes form to fit its environment. Since the tables were stabilized using solely the weight of other tables, a strong material was needed for their tops: Caesarstone, a substance that is made by resolidifying natural crystals that have been crushed into powder. When huddled together, the small tables created an elevated surface, which resembled a series of stepping stones floating effortlessly above the ground.

Mirror-Chair

Glas Italia

A mirror and a chair combined into one piece of furniture – one in which the latter component can be used as a seat or serve as a ledge for objects. Progressively darker frosted gradations at the junction of its two constituent parts, and on the chair seat itself, emphasize the fusion of these two functional elements while softening the chair's reflectivity, giving it a stronger physical presence and avoiding it blending into the mirror too much.

Rotating-Glass Shelf

Dilmos Milano, Milan

A storage unit whose 'doors' are formed from circular glass sheets. Their distinct design is derived from the historical practice of flattening and spinning blown 'crown glass' to produce thin, glazed discs. The dappled texture that results from this artisanal handmade process gently warps the perspective of the objects placed on the shelves behind the glass. The holes in the centre of the discs become finger-sized 'handles' by which to roll the discs left and right to gain access to the shelves.

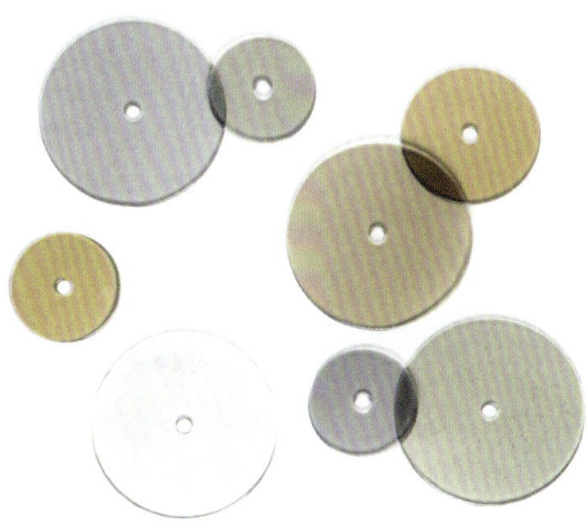

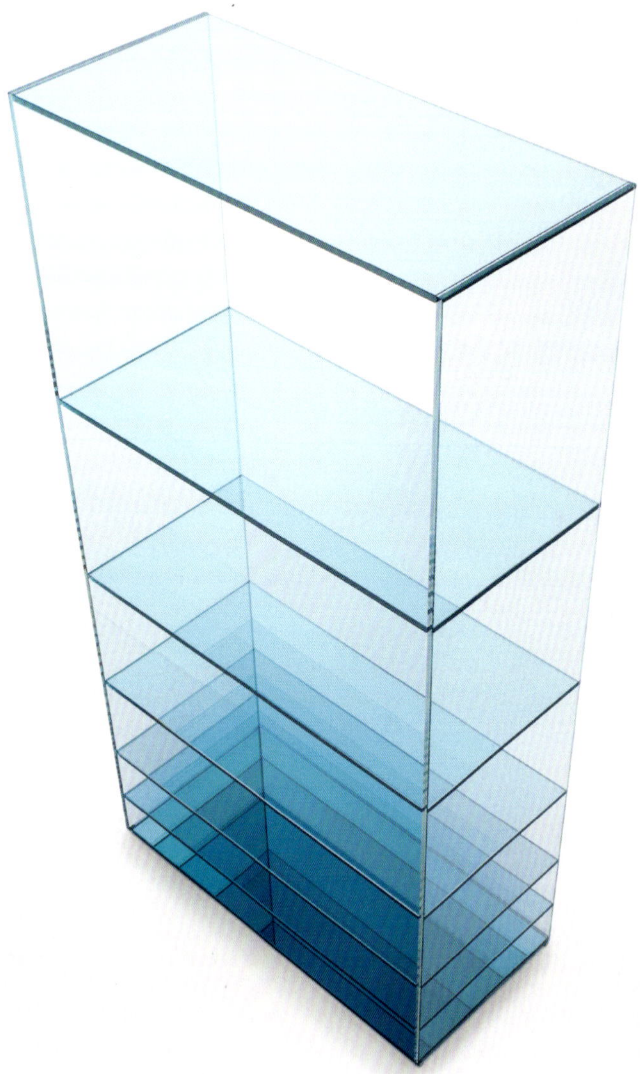

Deep-Sea

Glas Italia

It is possible to tint glass by melting a layer of transparent coloured film onto its surface. The Deep-Sea shelving unit builds on this technique by experimenting with the combination and number of film layers so that the shade of each successive glass sheet grows deeper. The space between the sheets was also progressively narrowed, and the base lined with a mirror, to further emphasize these gradations. The result was a piece of furniture with depth: transparent from any angle and drawing the viewer in, as though gazing at the surface of the sea.

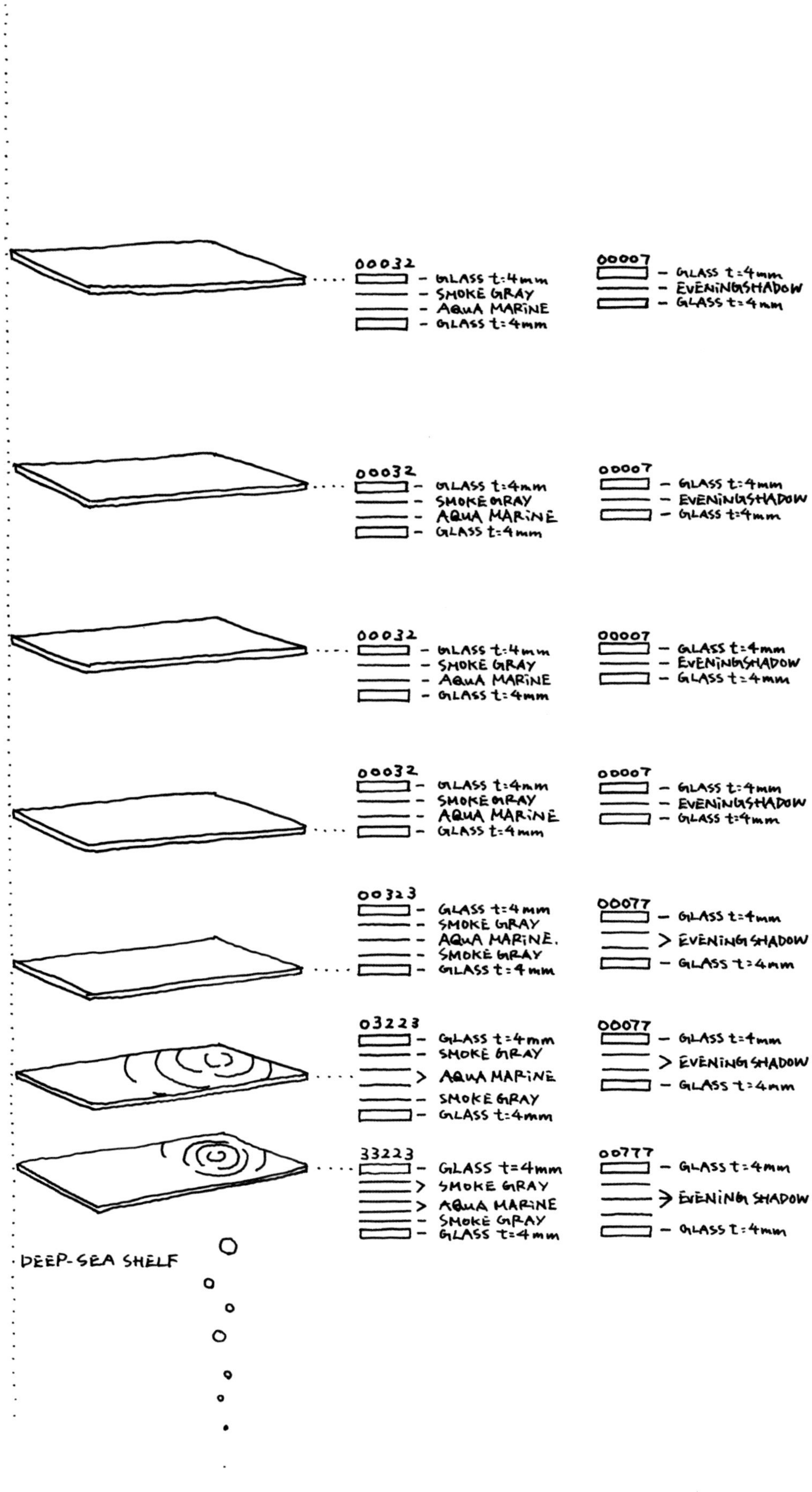
00032
GLASS t:4mm
SMOKE GRAY
AQUA MARINE
GLASS t:4mm
00007
GLASS t:4mm
EVENING SHADOW
GLASS t:4mm
00323
GLASS t:4mm
SMOKE GRAY
AQUA MARINE.
SMOKE GRAY
GLASS t:4mm
00077
03223
33223
00777
DEEP-SEA SHELF

343

Spicebox Office

Spicebox

An office-space design for digital agency Spicebox. The firm's name symbolizes its ability to deliver surprises and delight, like a variety of stimuli that come tumbling out of a box of exotic spices.

Accordingly, seven box-shaped meeting rooms of varying sizes and finishes were placed around the office, all of which are entered by a wall that has been 'flipped' open. Each box is opened in a slightly different way, subtly connecting interior and exterior and shaping sight lines to create a spatial experience in which the various, freestanding spaces reveal themselves as the visitor walks around the larger, outer room. The flipped-open entrances turn the area around each box into a semi-open meeting hub or breakout space, and functional elements that don't need to be on display – like the photocopier, waste bins and refrigerator – can be hidden behind the boxes. The idea was to make not only the boxes' interiors but also the area around them into an active office environment where people could easily work and interact.

Theory

Theory

An interior design for US fashion brand Theory that was rolled out across a number of stores worldwide. Intending to reflect Theory's fusion of functionality with casual, contemporary trends, the design adopted a New York loft-style aesthetic and materials, while adding a new concept: the flow of people. A circulation plan, such as an urban planner might use, ensures a natural flow of customers into and around the premises. One store boasts a 'boulevard' following on from the street crossing outside; another has access on both outward-facing walls of its corner entrance, with a curving 'short cut' between them.

Lampshower

Axor

Asked to combine a lamp with a shower, this design, unconventionally, chose to add the latter to the former. The resulting product serves as a transparent lamp when the shower is not in use. When functioning as a shower, however, light spills out from within the water jets, turning the ensemble into a luminous object. Three versions were created: pendant, wall-mounted and floor-standing. The wall and floor designs are not confined to the bathroom but can be moved around the house, or even taken outside, bringing showering to unexpected spaces of the home.

Paper-Brick

Pen

A set of flat paper blocks designed for an insert in *Pen*, a Japanese lifestyle magazine. They are printed with a wooden texture in three different shades, which gives the illusion of depth. When combined meticulously, the blocks can form a facsimile of one large, perfectly rendered timber cube; when scattered and mixed, they can – since they can't collapse – create 'impossible' forms that defy gravity. Inspired by the experience of digitally making 3D forms on 2D screens, this toy aims to translate the concept of 3D-CAD into analogue form.

Slice
Nava Design

The number and marking incisions on the face of the Slice wristwatch continue round its sides, giving the impression that it has been cut from a cylindrical piece of metal at regular intervals. As its hands sweep along their preordained paths, always at the same pace and speed, the watch reminds wearers of its function as a tool that 'carves out' time.

Cape
Offecct

A chair combining a metal-pipe frame with a seat and back made from two pieces of upholstered plywood, which are partially bonded on three sides, in the manner of a fabric or leather bag. Once this 'bag' is casually tossed over the frame, it looks like the chair is wearing a cape. There is even a subtle overhang at the base of the backrest, which reinforces this novel image.

Roof and Mushrooms

Kyoto University of Art and Design

A structure made in collaboration with architect Ryue Nishizawa. Situated on a steep, wooded hillside in Kyoto's University of Art and Design campus, the construction uses a single roof, inclined with the site's slope, to form a sweeping canopy that reminds visitors of walking outdoors under thick tree cover.

Many of Nishizawa's buildings feel like open fields, and their furniture like wildflowers, blurring interior and exterior boundaries. For this hillside structure, mushroom-shaped stools clustered around pillars, or in nooks by walls and staircases, evoking the growth of funghi in the wild. Some stools even grew from the railings, creating an ambiguous relationship between furniture and architecture.

Patchwork-Glass

Lasvit

A collection of objects combining the fine cut-glass techniques particular to Bohemian crystal with an ancient production method for sheet glass, in which the material is blown into the form of a cylinder, cut, opened up and flattened.

A variety of objects already decorated with traditional cut-glass patterns were reheated, then sliced open and reattached to each other to create one large object. As a manner of making, the process was like sewing together animal hides, or piecing together small fragments of cloth to create a great patchwork quilt.

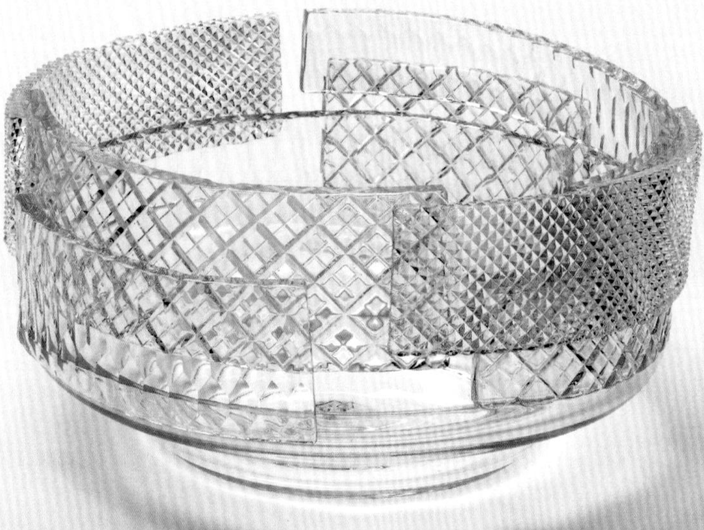

PATTERN A
PATTERN B
PATTERN C
PATTERN D
PATTERN E
PATTERN F
PATTERN G
PATTERN H

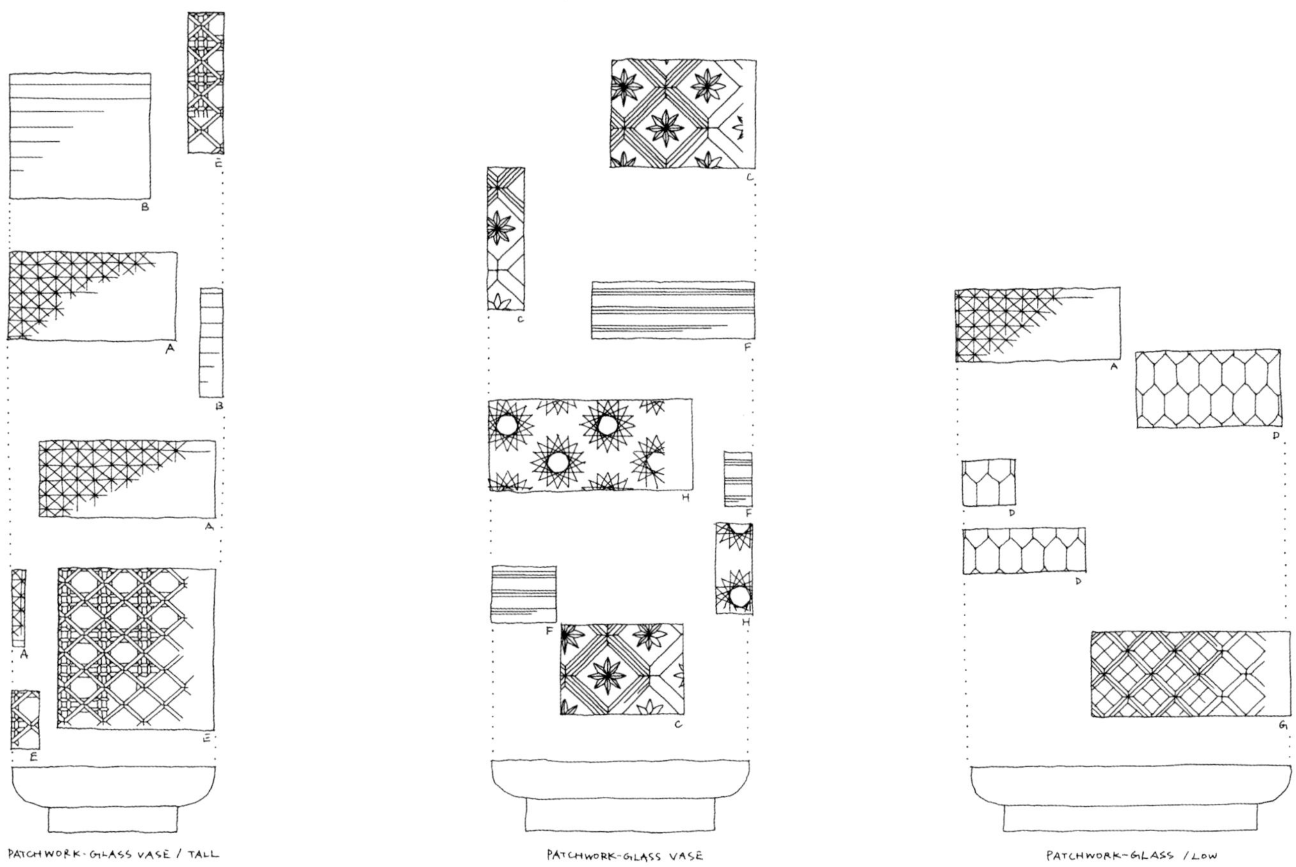
PATCHWORK-GLASS VASE / TALL
PATCHWORK-GLASS VASE
PATCHWORK-GLASS / LOW

Sekki

Luckywood

A three-piece dessert-cutlery set whose warped and crooked outlines are reminiscent of prehistoric flint implements. The distinctive thickness and weight of the pieces, similar to those of stoneware, required meticulous handwork and seven rounds of metal pressing, rather than the usual once. The concave areas of the utensils' surfaces were sandblasted for a matte finish and the remainder polished to a mirror-like smoothness and sheen, giving the impression that the cutlery was carved from stone.

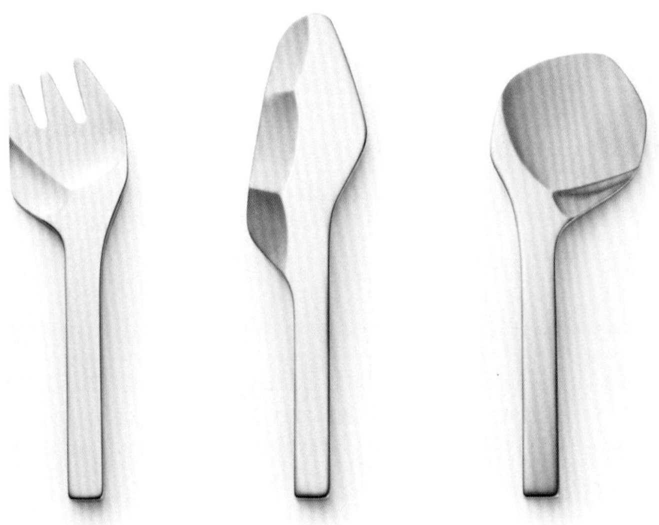

Rassen

Hashikura Matsukan

For four centuries, the town of Obama in Fukui Prefecture has manufactured lacquered chopsticks. They have been recognized as the most durable and beautiful examples of their kind since the seventeenth century, when they became known as *Wakasa-nuri* (denoting lacquerware patterned with various materials).

The Rassen chopsticks were designed for Hashikura Matsukan, manufacturers who continue Obama's traditional techniques. Chopsticks ordinarily come in pairs, but these form a single unit that is separated into two for eating, then rejoined when not in use. The chopsticks were created using both artisanal hand skills and multi-axis CNC milling.

Uneven-Oke Bucket

Nakagawa Mokkougei

A champagne cooler whose slats have been left at uneven heights, hinting at how it was made. While a seemingly minor design detail, this decision resulted in a complex production process: each cooler must be assembled once, then taken apart, its surfaces finished then reassembled. Unlike ordinary Japanese wooden buckets, which have a flat bottom and straight sides, the cooler's base is curved for a softer look and feel. Its hoops have been sandblasted and then powder-painted in order to dull their metallic gleam. The coolers are made from traditional *hinoki* (Japanese cypress) which is preserved with a urethane treatment to minimize warping due to UV rays, dirt and moisture. The combination of traditional Japanese cooperage craftsmanship and modern techniques gives this champagne cooler a special resonance.

Heel

Moroso

Inspired by the spiked heels on a pair of stilettos, the line of the back legs and their connection to the backrest give this chair the silhouette of a high-heeled shoe. It is made from timber, with a hollow-section backrest, and lacquered in black.

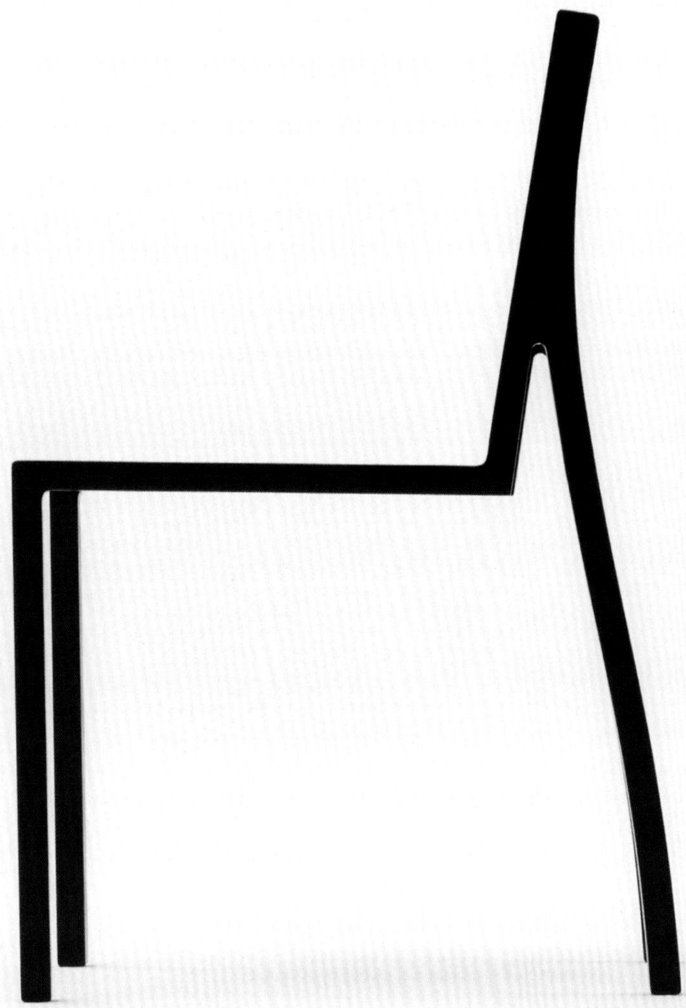

Basket-Lamp

Kanaami-Tsuji

Wire netting *(kana-ami)* cooking utensils have been employed in the city of Kyoto to create its delicate, exquisite cuisine for over a thousand years. Made by artisans, their designs are said to have originated from the beauty of the shadow cast by the netting used to lift blocks of tofu from boiling water.

Created in collaboration with Kanaami-Tsuji, a Kyoto-based wire-netting firm, Basket-Lamp preserves the craft's traditions and develops them for contemporary living. The design allows the handcrafted shade's woven pattern to reflect on the ceiling, and its powder-paint finish further accentuates the lamp's distinctive silhouette.

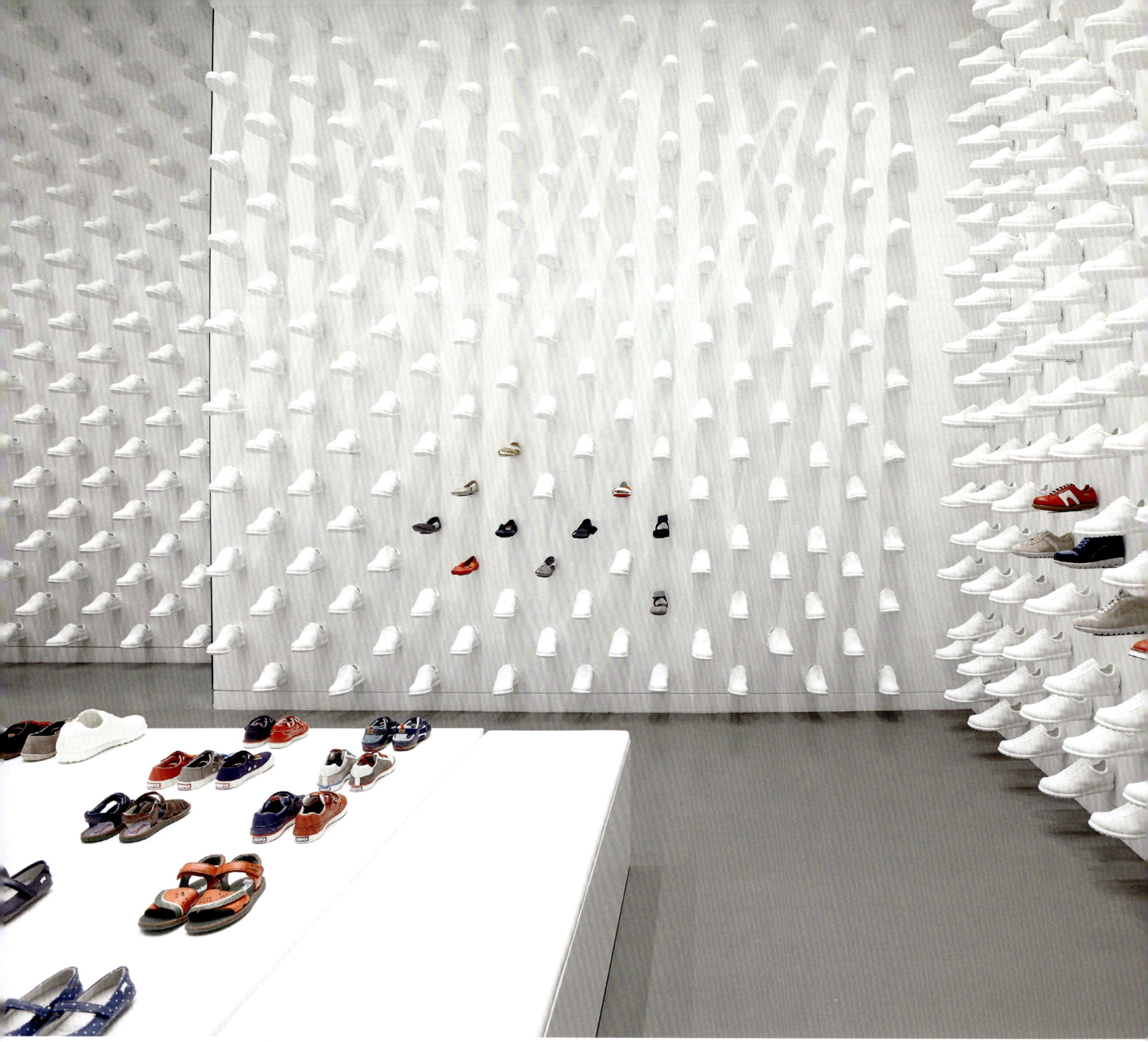

Camper 2nd Concept

Camper

Following on from nendo's first Camper concept of 'shoes walking freely through the air', which was developed for the firm's smaller retail outlets, a second design was developed for its larger spaces in Madrid and New York

Due to the increase in scale, the new concept faced a consistent challenge: sufficiently utilizing the upper areas of their high-ceilinged spaces. To take advantage of this potential, the new approach involved making white resin models of the Pelotas shoe, a Camper classic, and then decorating the walls with them, floor to ceiling, in a regular grid pattern. This added to the retail space's perceived volume as well as evoking the orderly atmosphere of a storeroom. The shoes catch light pouring through the store windows and cast shadows on the walls, giving the space a three-dimensional texture. The result is a shop interior that showcases two kinds of shoe: one as a conventional product, the other as an interior-design feature.

367

Chocolate-Paint

Hunter Confectionery

Tubes in a paintbox contain a variety of colours – and these chocolates, likewise, a variety of flavoured syrups. Presented like a set of oil paints, the chocolates' labels indicate each one's flavour, but also function as wrappers. The result is a design that combines the childhood excitement of starting a new box of paints and the thrill of opening a box of chocolates.

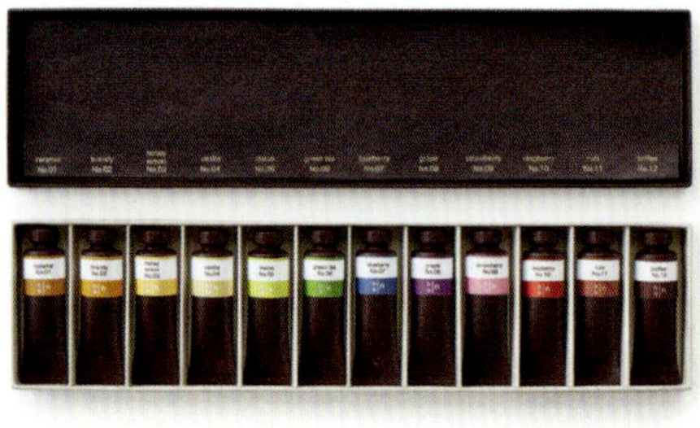

Lime-Pocket
Bombay Sapphire

An accessory designed for Bombay Sapphire, the gin known for its sapphire-blue bottles. It is usual to add a twist of lemon or lime to cocktails like a gin and tonic and a Tom Collins. To accommodate the required citrus wedge, a custom-made silicone 'jacket' was designed to keep fingers clean and dry, and to provide a neat holder for the slice once squeezed. The jacket is the appropriate shade of blue and bears Bombay Sapphire's logo in order to heighten brand visibility when the slit wedge is hooked onto the edge of the cocktail glass.

Heads or Tails
Pen

A three-piece dog-accessory collection consisting of a bed, dish and toy – each with multiple uses and forms. The bed becomes an igloo-like hut when dogs burrow inside it, and a cushion when they lie on top of it. The similarly faceted ceramic dishes are reversible, with a shallow dish for food on one side and a deeper bowl for water on the other. The silicone-rubber, dumbbell-like 'bone' can be reshaped into a ball by folding in its two ends.

As a result of searching for a structure that could maintain its stability in two different forms, the collection is constructed of triangular panels connected in a polygonal mesh.

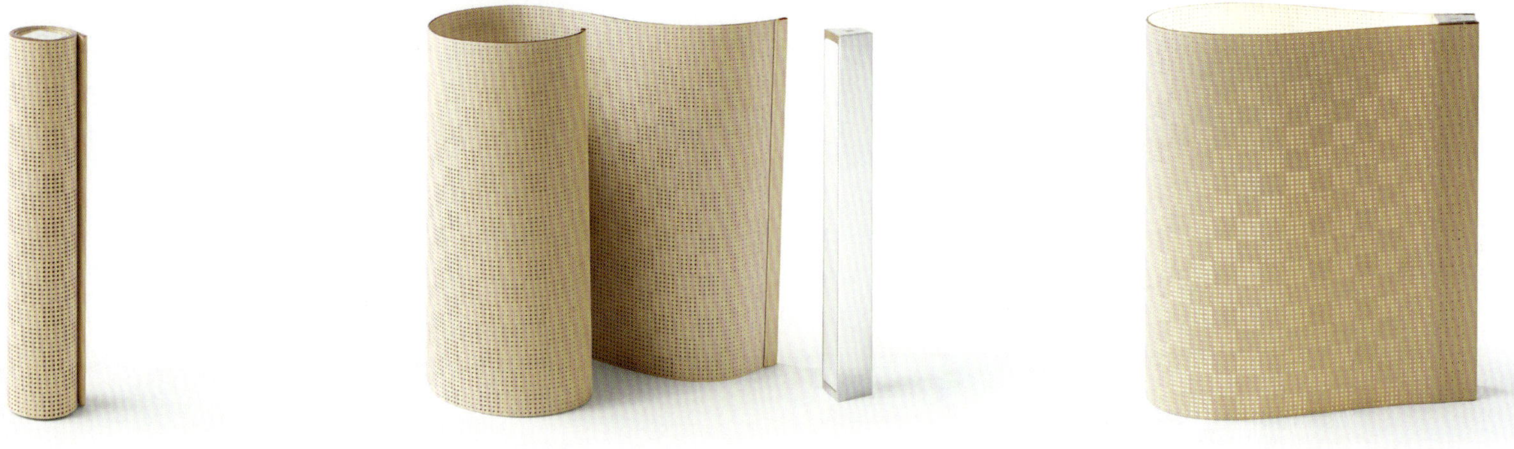

Surface

Louis Vuitton

A lamp designed for Louis Vuitton's first ever furniture collection. Travel has been the company's theme since its establishment in 1854, so the objective was to make a portable design. Inspired by the brand's workshop, where stacks of rolled up leather wait to be transformed into bags and other accessories, the lamp utilizes a single piece of leather that can be rolled up for transport. When in use, the leather is fixed in place by magnets to function as a lampshade. The material is punched with holes, creating a Damier pattern for light to shine through, showcasing Louis Vuitton's craftsmanship and trademark design.

Compolux

Sogo & Seibu

A design for the display areas in the women's clothing department of the Seibu store in Tokyo's Shibuya district. Multiple brands had to be presented together in a unified environment, but each needed to be gently distinguished from the others. The brands' lineups and product arrangements change at a dizzying pace, so the fixtures needed to be easily modifiable.

For the overall division of the space, Compolux took inspiration from the wrought-iron fences surrounding the parks, squares and other green spaces in European cities. Screens were created based on these fences, and suspended from the ceiling as hanger racks for the clothes. The screens are easy to remove and re-locate, and have built-in lighting to illuminate the merchandise. They come in six different patterns in order to give each brand a distinctive look. Further building on the urban-park aesthetic, shelves that extended from the screens, low floor fixtures inspired by park benches and a circular seat reminiscent of a fountain's edge were also included in the display. These designs created a free, lighthearted environment, similar to the experience of strolling in a park.

374

Ofon

Kokuyo

An office-furniture collection for small workspaces needing to rapidly modify layouts in response to fluctuating employee numbers. It is expandable and multifunctional, using a joint that can be screwed and unscrewed with a coin, rather than requiring tools. Unused shelves serve as 'legs' for cabinet-type desks; upholstered panels function as cupboard doors and also as partitions. The two parts of the 'coin joint' lie recessed when the furniture is not in use, but spring out with a finger-touch when modifications to the layout are needed.

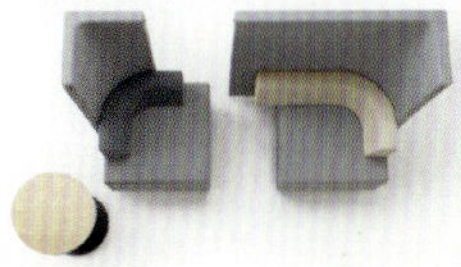

Brackets

Kokuyo

An office-use unit sofa designed to create a communication space by 'bracketing' its inhabitants like parentheses. Users can freely combine the seven sofa types and four table variations to achieve a customized seating area balancing openness and privacy, concentration and relaxation, and individuals and groups. The sofas are composed of sharp outer lines and softer inner curves, so that they respect the mood of the office environment while providing for user comfort. The height of the backrests, which are relatively tall at 1.4 metres (5 feet), provides visual privacy and dampens the surrounding noise, allowing for focused conversations.

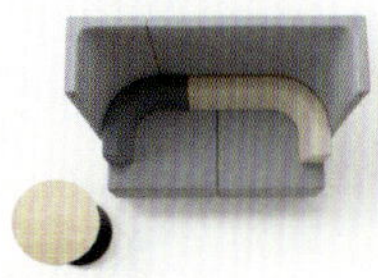

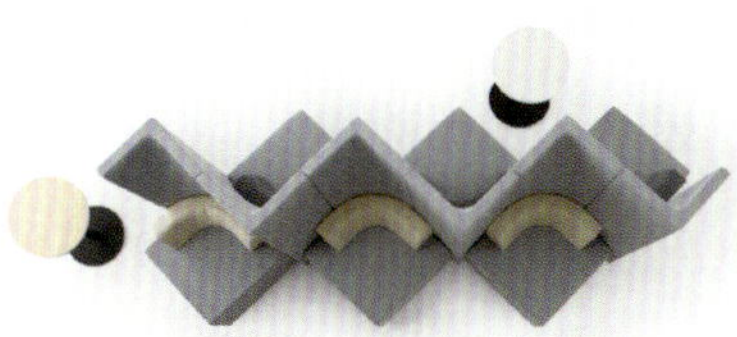

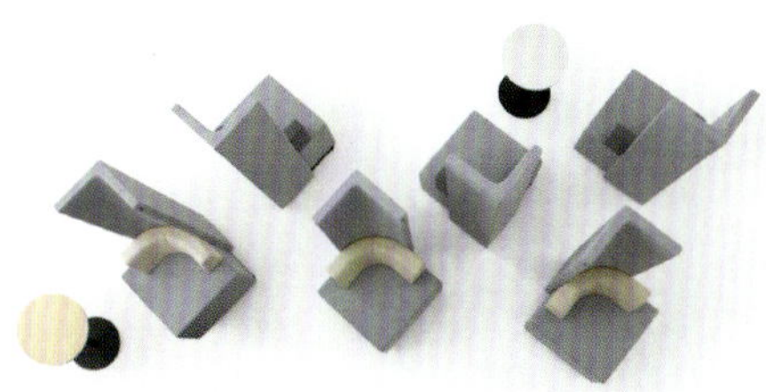

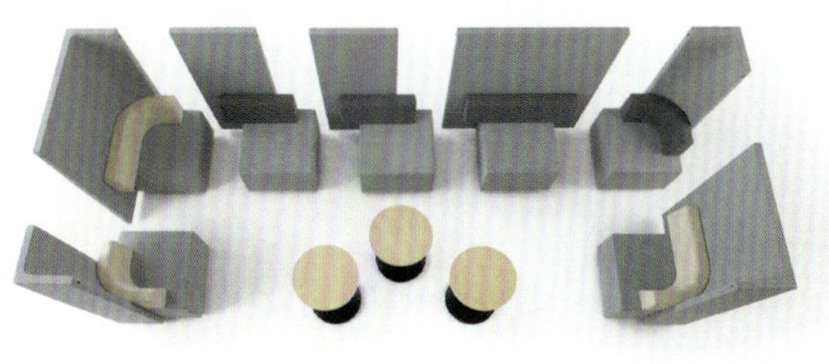

Nichetto=Nendo

Milan Design Week

A seven-piece furniture collection created in collaboration with Italian designer Luca Nichetto. In a process similar to the Japanese poetry genre known as *tanka* (short poem), in which the opening three lines generate a two-line response, this partnership took turns devising basic concepts that were then passed over to the other party to use as inspiration, and to further expand and develop. Working collaboratively in this way, the design process was like a jam session, with each participant taking turns improvising melodies on top of a shared rhythm until a 'song' emerged.

Coloured-Pencil Table

Saint-Etienne International Design Biennial

A project using *Udukuri*, a traditional technique in which the soft parts of wood are carved away so that the curves and lines of the harder areas stand out. The method was applied to cypress boards, which were then covered with paper and the resulting relief pattern traced in multiple shades of coloured pencil in order to transfer the image of the wood grain on to the paper. The result was a distinctive juxtaposition between the familiar appearance of the wood, the finish of the paper and the naturally uneven outcome of the hand colouring process.

379

2014

Softer than Steel

Desalto

A section of the Softer than Steel bench has been pushed up to form a gentle hump that functions as its backrest; poles wrap around each other to become a coat stand; wall-mounted shelves curl up at their base, like a sheet of paper; two opposite legs of a chair extend upwards and join in a high back-loop reminiscent of a single paintbrush stroke; nesting tables feature smoothly curved cut-outs in their surfaces, allowing the legs of the taller tables to pass through them and the set to be combined.

By adding curved, bent and wrapped details to steel sheets and rods, the ordinarily hard material gains new functionality and a light, flexible feel – as though the metal surfaces have actually become paper or cloth.

388

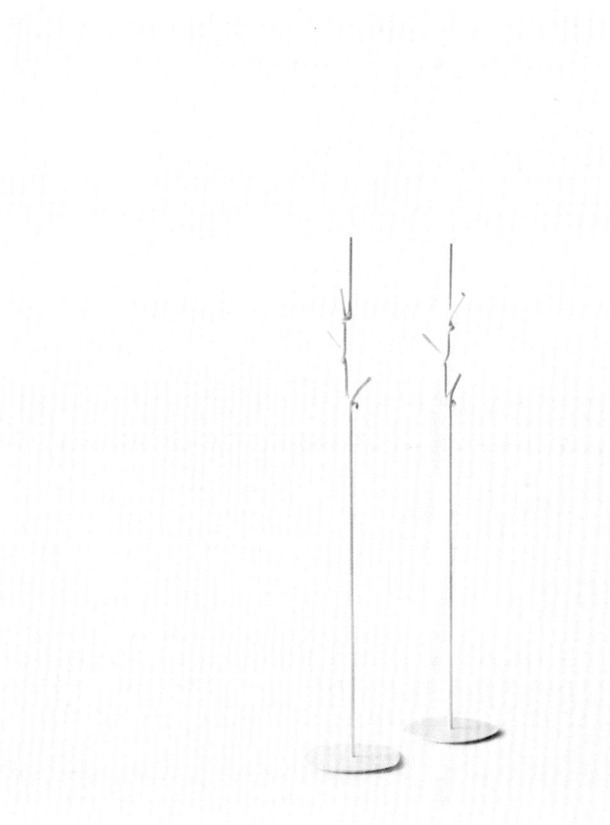

Beige 2nd Concept

Onward Kashiyama

Fashion brand Beige wanted its multifunctional Tokyo space to serve as a library and art gallery, in addition to retail clothing. Dividing the store into separate rooms was deemed unsuitable for the already small area; instead, the space was stratified vertically. A thin metal beam, installed 2 metres (6½ feet) above floor level, created shelf for books. Movable clothing and bag hangers were suspended from this shelf, and box-shaped fixtures dotted the floor below – usable as stepstools to reach the books or plinths for exhibiting artworks and accessories.

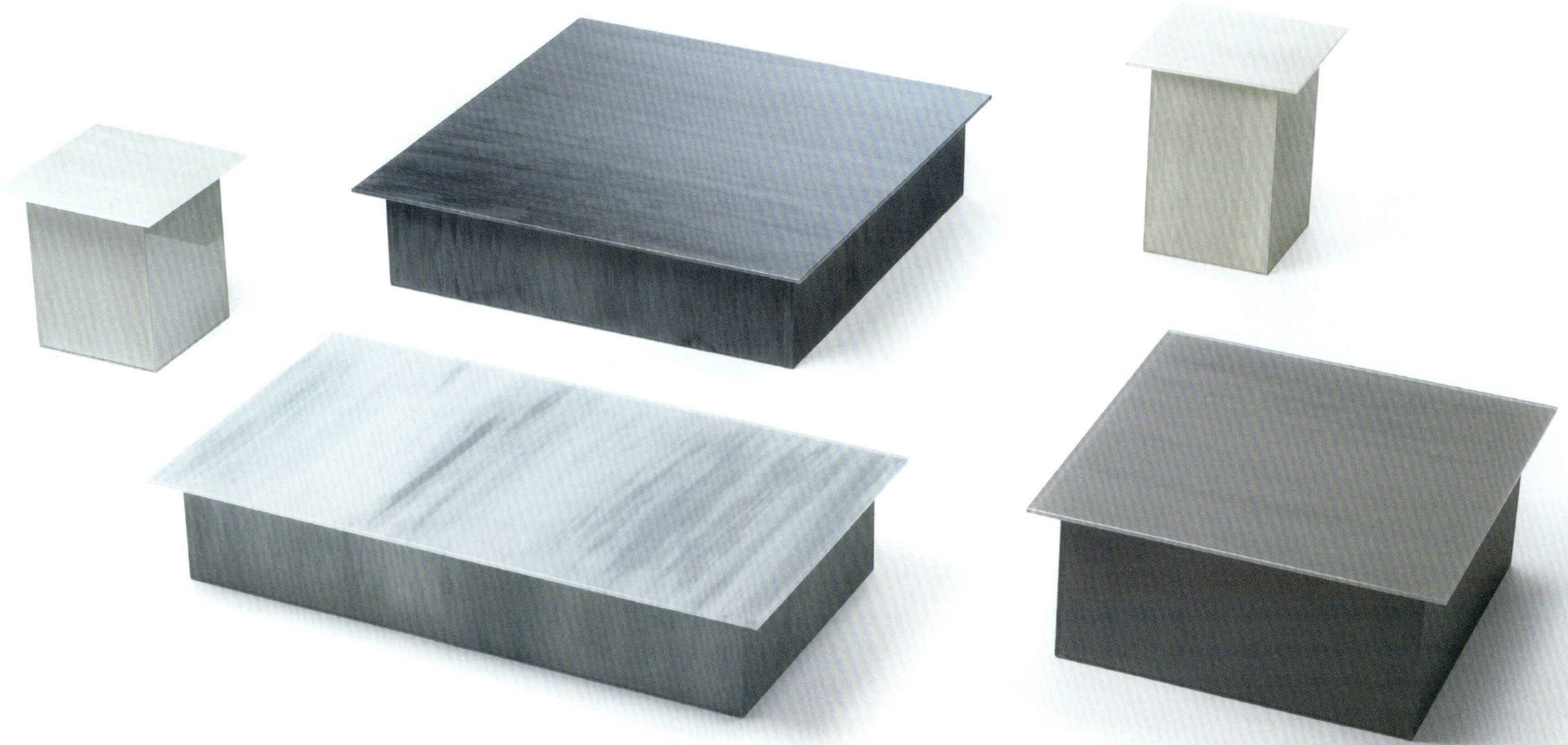

Brushstroke

Glas Italia

A furniture collection – including a side table, console, bench and dining table – coloured with paint applied using two different methods. The first layer was brushed onto the transparent glass surface, then another (of a different colour) was sprayed over the top. Mixing colours in this way, combined with the unevenness of the brushmarks, produced a natural-looking effect, similar to wood or stone. The juxtaposition of the glossiness and smoothness inherent in glass with the imperfect finish of the hand-applied paint created a series of unique material appearances.

Harcourt-Chessboard

Baccarat

A chess set made from Harcourt glass, the crystal pioneered in 1841 by the famous French Baccarat brand. To create each chess piece, six elements from the glassware's basic structure were isolated to represent the figure's distinctive characteristics. Crafted from both transparent and midnight-blue crystal, whose multi-faceted nature makes it incredibly versatile, the pieces appearance changes dramatically even through subtle variations in the position and angle of cuts. The squares of the chessboard are delineated by Plexiglas panes, which diffuse light and create a shimmering quality.

In combining all of these elements, the design aims to bring the conviviality of a shared drink to a game of chess.

SU
Emeco

Aiming to embody the spirit of Emeco's legendary 1006 Navy Chair, which has been manufactured for the US ship and submarine fleet since 1944, the SU stool is versatile in its materials, dimensions and form. Its seat, which replicates the curves and indentations of its inspiration for comfort, can be easily removed with a single coin and its legs are stackable. The seat is available in three low-environmental-impact materials – recycled PET resin; reclaimed oak; and a lightweight, energy-efficient 'eco-concrete', made from recycled glass bottles and calcium sulfo-aluminate cement (CSA). The legs come in oak and black-painted or natural aluminium. *Su* in Japanese means basic, simple and pure. With all unnecessary elements removed from its design, the stool aims to live up to its name.

Eigruob
Kartell

To commemorate the tenth anniversary of its iconic Bourgie table lamp, Italian plastic-furniture specialist Kartell invited ten designers associated with the company to create tribute pieces for a special exhibition. This contribution worked with two of the lamp's most distinctive characteristics – its silhouette and its transparency – rather than altering the original design itself.

Eigruob was created by inverting and mirroring a 'negative' of Bourgie's profile. When two of the new lamps are lined up close to each other, the space between them forms an upside-down silhouette of the Bourgie lamp, in a manner reminiscent of the classic 'Rubin vase' *Gestalt*-theory image. Because this homage reverses both the lamp's outline and our regular sense of 'up and down', it was named 'Eigruob'.

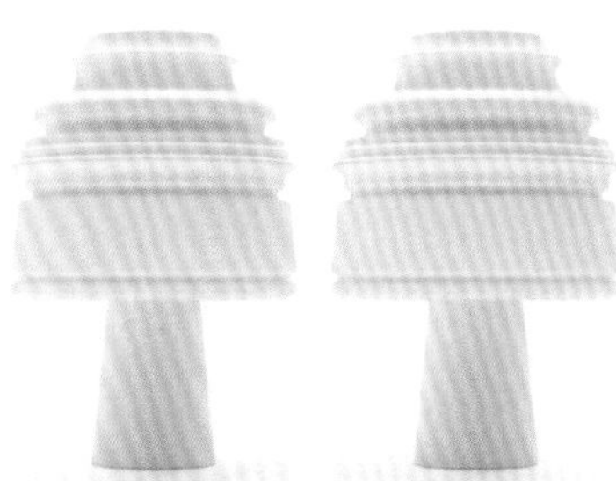

Pooh-Glassware

Walt Disney

A collection of Winnie-the-Pooh-inspired glassware for Walt Disney, Japan. Rather than focusing on the glassware itself, the design concentrates on a series of accessories. The basic glass, bottle and jar shapes are complemented by silicone lids and bases that recreate some of Pooh's best-known escapades: the lids show Pooh climbing trees, putting his head in a jar of honey and hanging from a balloon; the bases or 'coasters' depict the door to Pooh's house and the clock in his room, honey on a branch and Pooh stuck in a hole.

Pooh-Table

Walt Disney

A collection of tables, also for Walt Disney, Japan, based on AA Milne's children's character Winnie-the-Pooh and his friends. To reflect the Pooh stories' setting in the fictional Hundred Acre Wood, the tables are made of beech wood and come in sizes and shapes intended to recall Milne's creations. Each character's distinctive clothing, hair and tail is subtly hinted at through coloured, knitted cladding on different parts of the tables.

Bi-Colour Washi

Taniguchi Aoya Washi

Washi is made from mulberry pulp collected and dried on screens, and peeled off in paper sheets. After experimenting with dyeing *washi* forms in two stages as they were being shaped, it became apparent that it was possible to create a lampshade whose inner and outer surfaces were different colours. Their dark or bright exteriors could be accompanied on the interior by the kind of light, reflective colour necessary for light diffusion – a first for traditional *washi* lampshade manufacture. The resulting shades can also be stacked, in order to keep distribution and storage costs to a minimum.

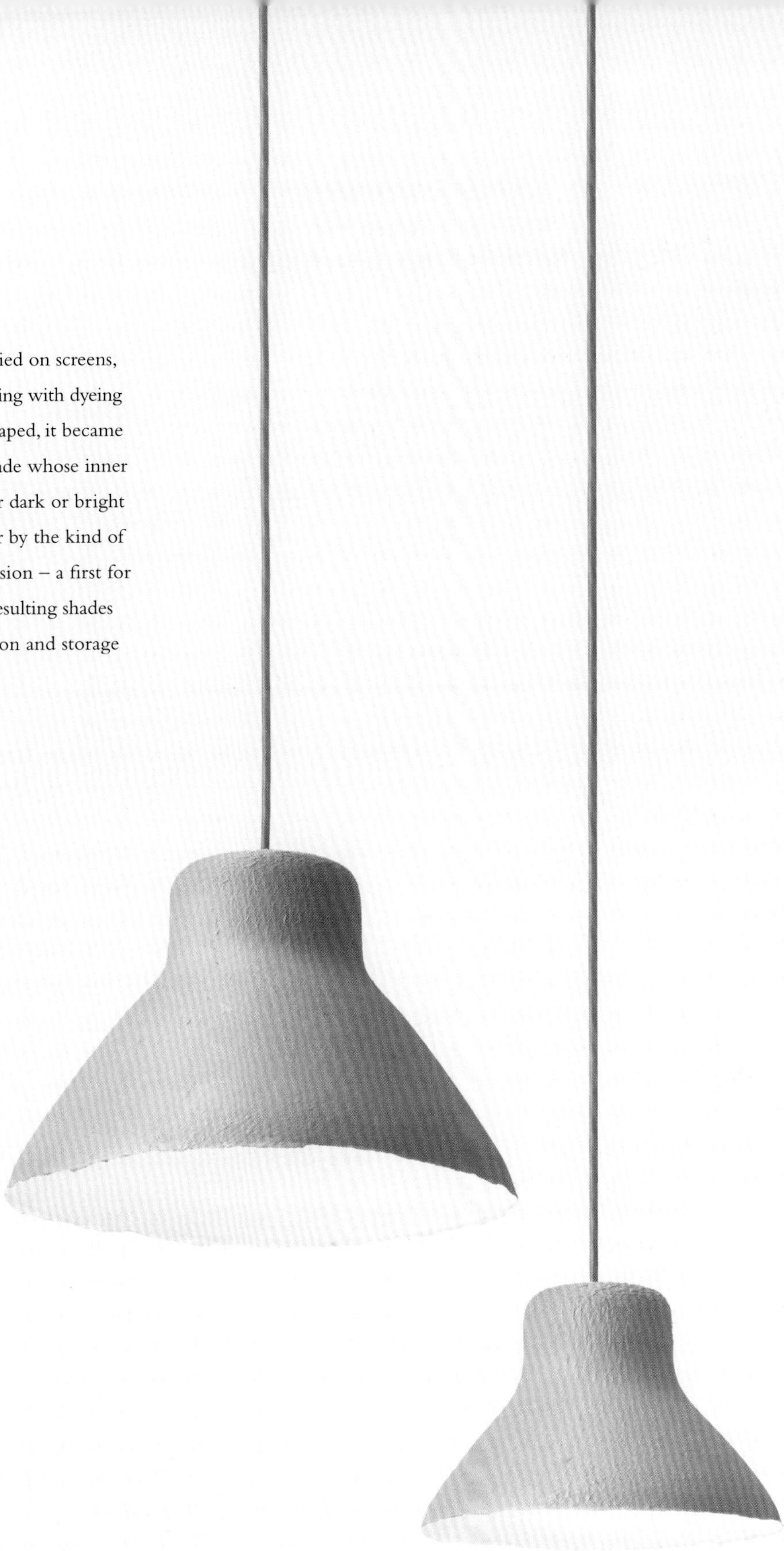

Space Dipped Shirts

Cos

The Cos 'look' combines simplicity with sophisticated details and carefully selected materials. One of the clothing brand's flagship products, its white shirt, starred in an installation at the 2014 Salone del Mobile in Milan.

Experimenting with the interplay between shirts and frames, the exhibition presented a series of striking sculptural arrangements. The ordered shirts are crisp, classic white until they 'fall' inside the empty steel-cube frames, at which point they take on colour as though the space itself has dyed them.

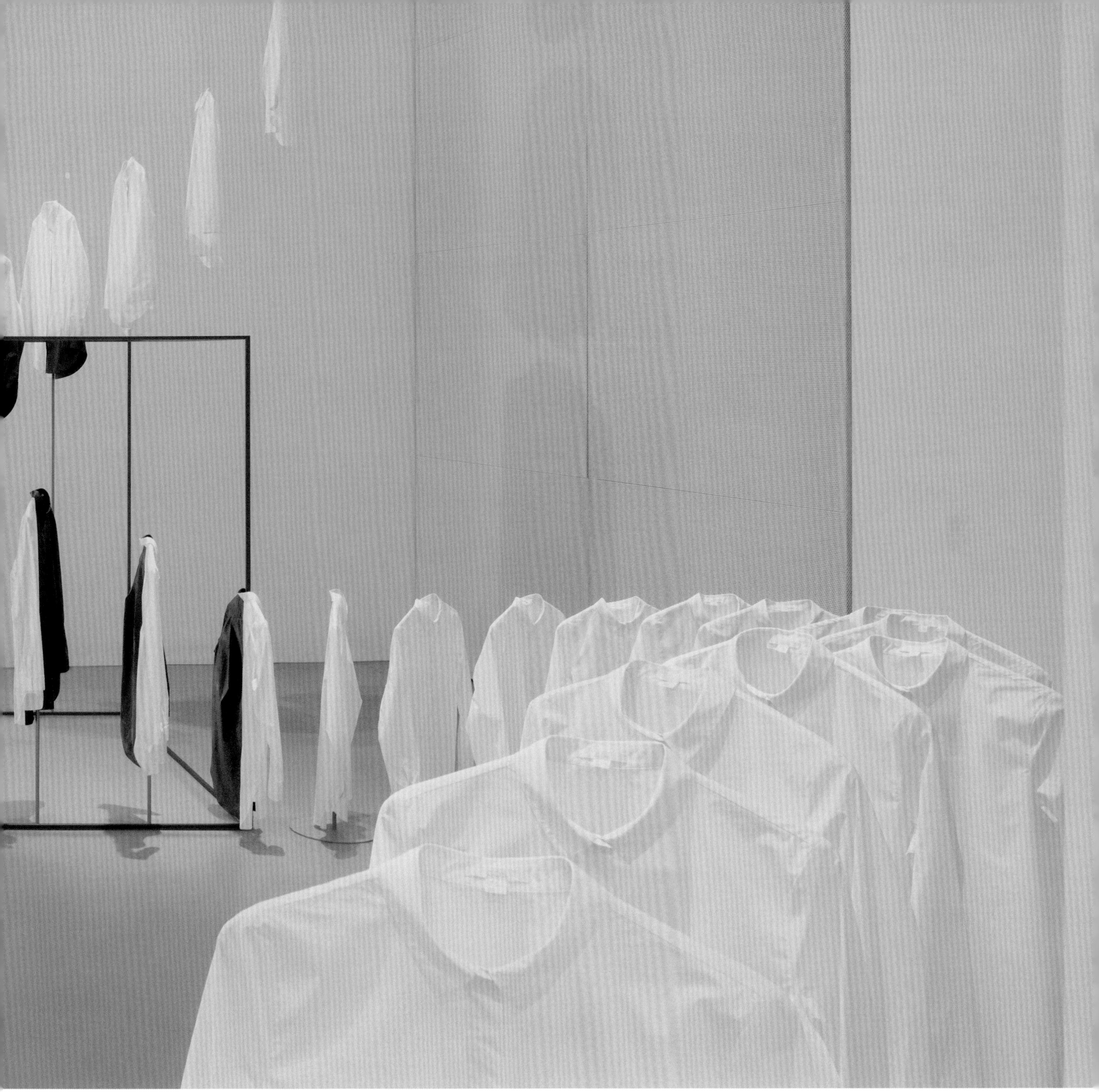

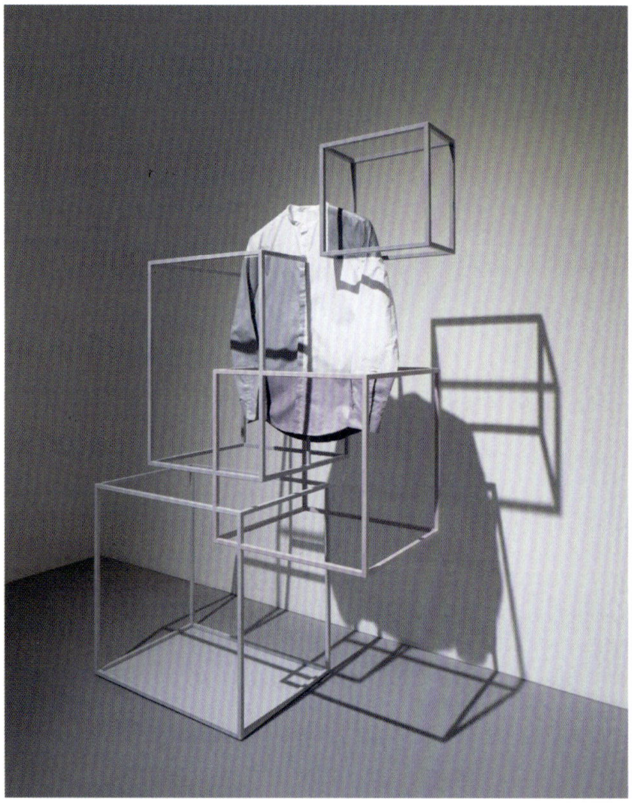

Print-Chair

Milan Design Week

A series of chairs whose surfaces combine two different patterns, created by printing naturally occurring configurations onto timber that already has a distinctive grain. Experiments with the degrees of combination resulted in a number of different outcomes. Some seats were layered with two different wood-grain patterns at the same scale, and others with enlarged, abstracted versions printed onto the existing one. On another design, the wood's original surface was scanned and then printed back onto the chair timber at a different angle. Other materials were also used: in one case, the seat base was replaced with OSB laminate board, and in another, a marble pattern was printed onto wood.

Thanks to the latest printing technology, continual fine adjustments could be made to the scale, density and colour of the printed patterns – showcasing the charm and attraction of these natural materials.

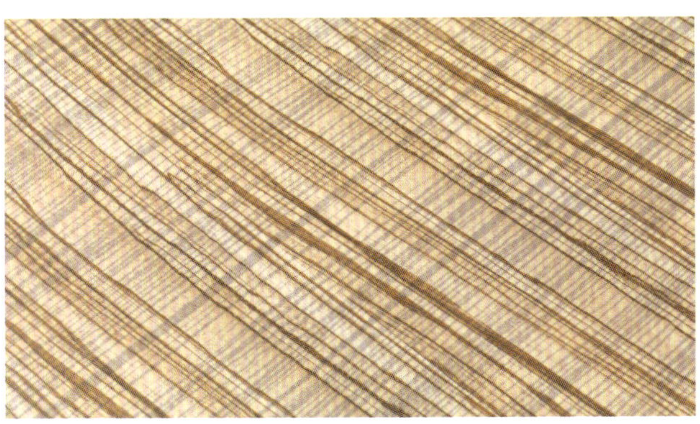

Hidden – Unveiling Japanese Design

Japan External Trade Organization

An exhibition project – organized by Japan's External Trade Organization (JETRO) and held at Singapore's National Design Centre – that involved the conceptualization, curation and design of the space. While overseas interest in Japanese design is growing, the understanding of its nuances remains superficial. The perception is that Japanese design is about reduction and minimalism. However, such 'removal' is actually a deft and skilful means of constructing a heirarchy, making one thing more attractive by concealing others. The exhibition was therefore focused on the attitude and intent lying at the heart of Japanese design, and to foreground the 'hidden' or overlooked elements of the objects on display.

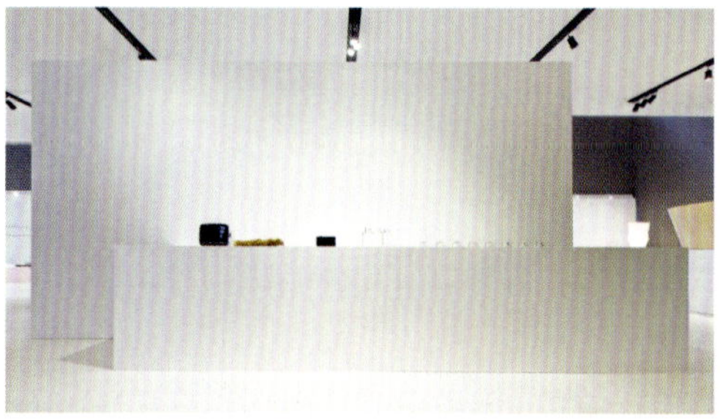

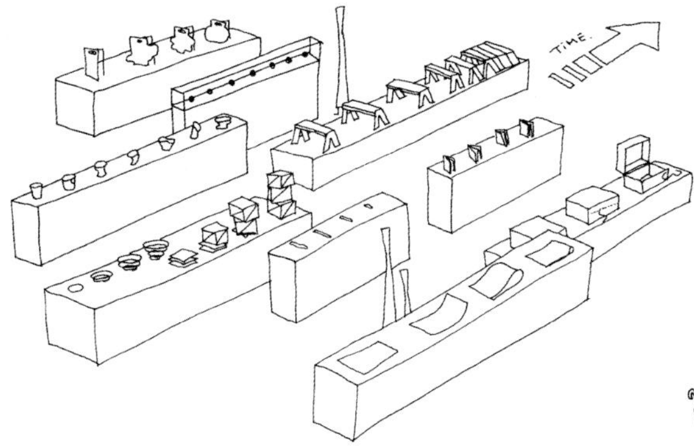
TIME

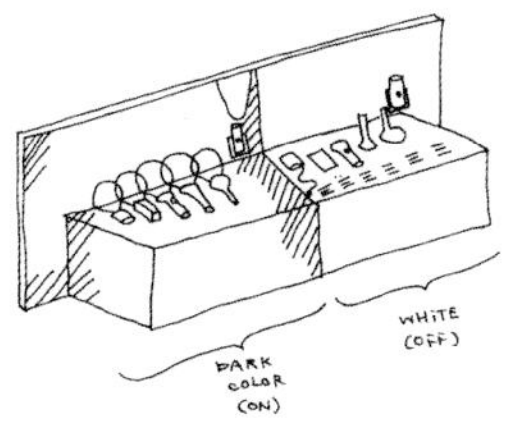
DARK COLOR (ON)
WHITE (OFF)

WOOD
PAPER
WOOD
RESIN
LAYERS OF METAL
URUSHI
POWDER
BAMBOO FIBER

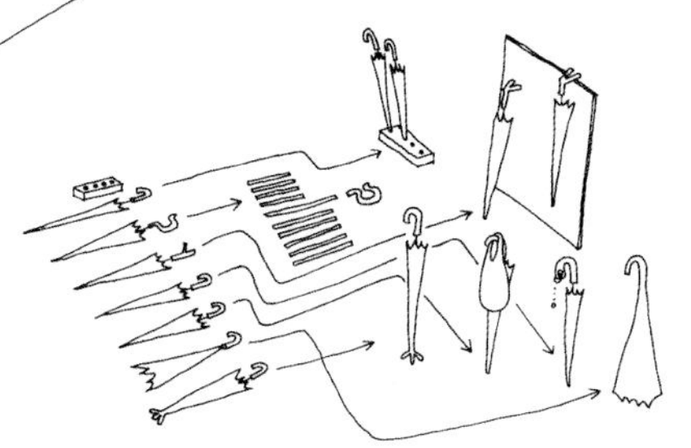

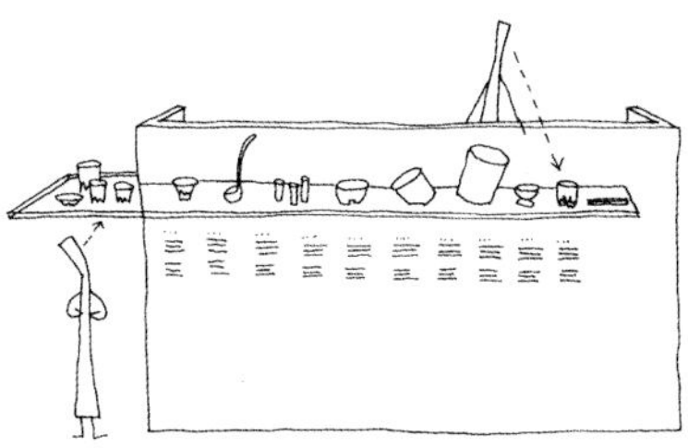

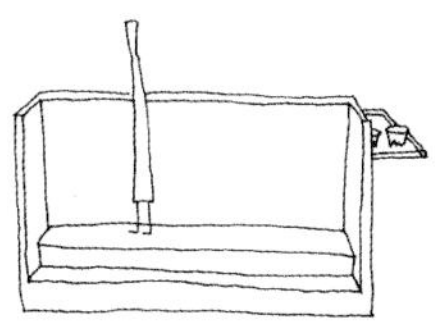

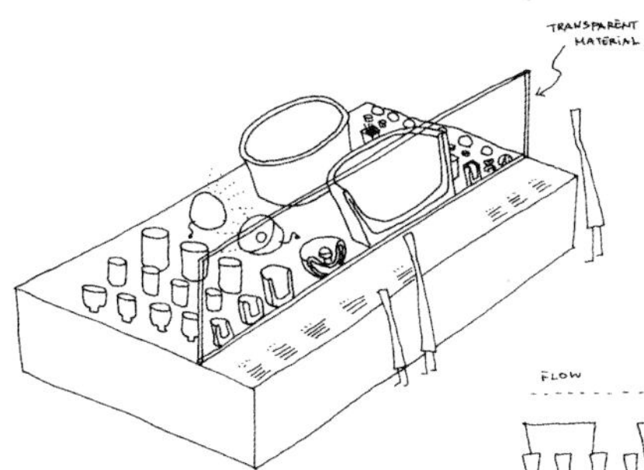
TRANSPARENT MATERIAL

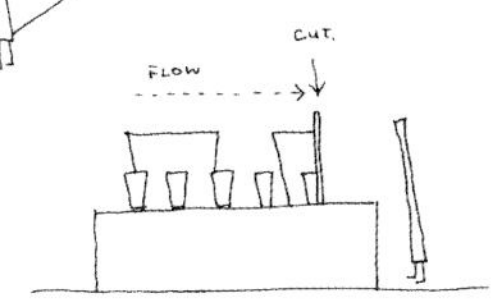
FLOW
CUT.

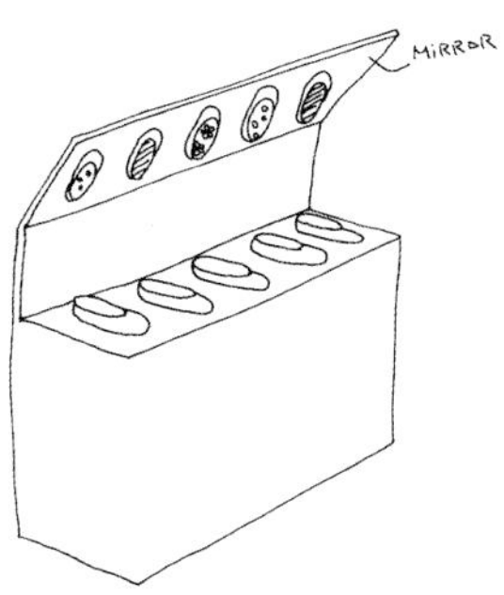
MIRROR

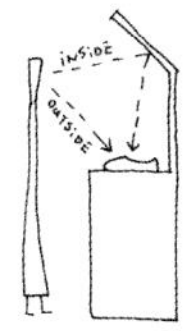
INSIDE
OUTSIDE

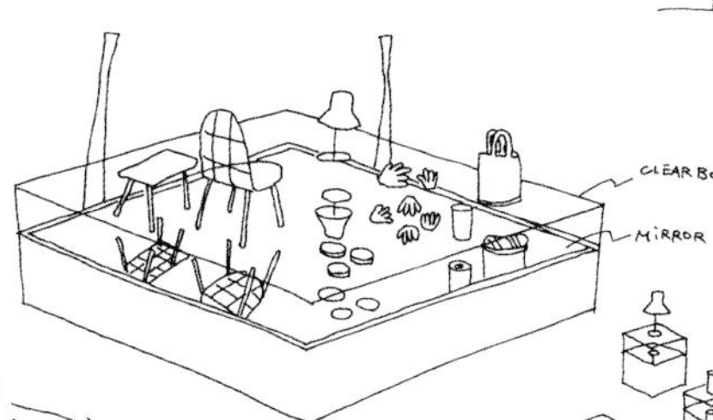
CLEAR BOX
MIRROR

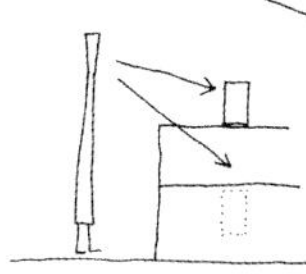

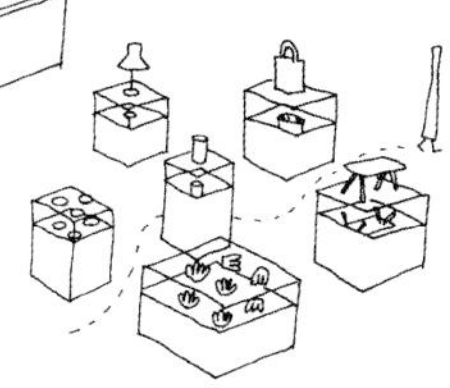

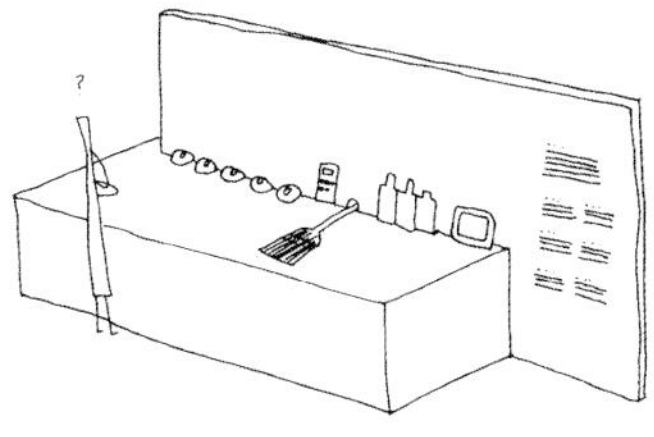

FRONT

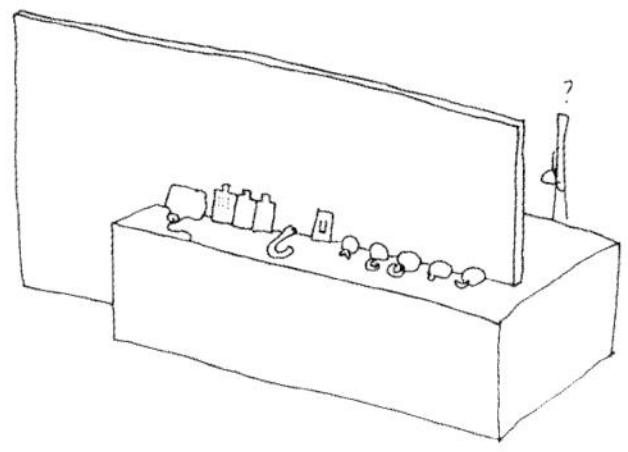

BACK

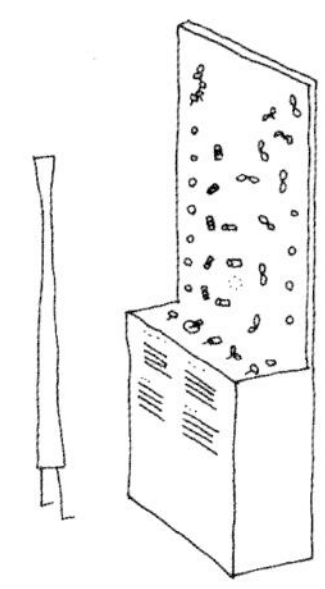

Camper Shoes

Camper

Camper's Beetle shoe is a lighter, more water-resistant version of its long-selling favourite: Peu Ideal. Exploring possibilities for taking the Beetle even further, the new design focused on Camper's iconic, stretchable shoelaces and knots. The laces were intertwined across the shoe's upper surface to provide comfort and wearability, while at the same time taking on a 'starring role', bringing new functionality and visual style to a design classic.

Camper Sunglasses

Camper

A collection of sunglasses for Camper. The distinctive design layered two lenses of different colours inside each frame, creating subtle, gradated shades of blue/black, brown/grey and red/green. Thanks to the skill of sunglass specialist Inui Lens, from Sabae, Fukui Prefecture, the lenses were combined perfectly with no space between for trapped air.

The cases employ Camper's signature soft leather, with a fastener made from a detachable glasses arm that is fixed to the case by a cord. A cleaning cloth imitates the lens shape, for a touch of Camper's characteristic playfulness.

Snow Dome

Grasse Tokyo

A spray-type room fragrance in the shape of a snow dome. It rests upside down on its base, which doubles as its cap, and, when lifted, reveals its pump mechanism. The 50-millilitre (1.8-fluid-ounce) domes are filled with tiny flowers, stars or heart shapes in polyester film. This discreet object disappears beautifully into the room when not in use, and comes in three fragrances: lavender and musk (flower shapes), 'forest' (stars) and ylang-ylang and cassis (hearts).

Draftsman

I Next GE

A collection of wristwatches inspired by the increment and numbering style of drafting instruments. The first in the series – Draftsmen 01.Scale – borrows the look of a clear, plastic ruler, with the hour and minute marks printed in the style of centimetre and millimetre increments. The marks also lie directly on the crystal rather than on the watch face itself, in order to emphasize the visual link with marks on the upper face of a clear ruler.

Continuing but also evolving the theme, Drafstmen 02.Scale has hour markers that were laser-cut through the crystal rather than printed onto its face. It was hoped that this might inspire wearers to become aware of the time that they measure for themselves – like the operations undertaken by a draftsperson – instead of simply following time as measured by others.

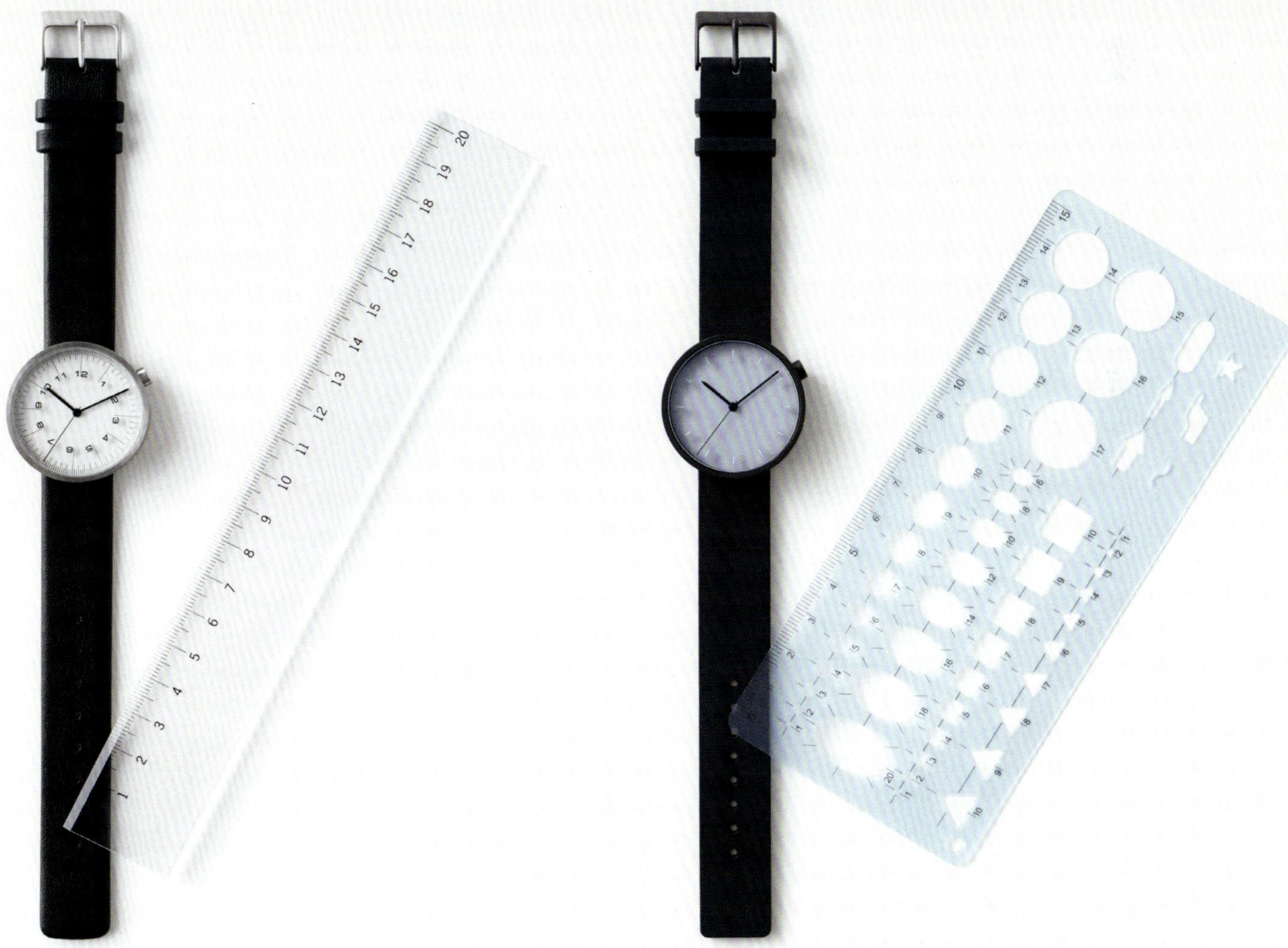

Shelf + Desk + Chair = Office

Kokuyo

While our demands on office space become ever more diverse, office furniture has not kept pace and remains oddly standardized. In contrast, this installation tried to return to the fundamental principles of the workplace by asking what, ultimately, an office is. The resulting possibilities were exhibited at the Orgatec 2014 trade fair, held in Cologne.

The identity of the collection was formed by breaking down the office into its smallest components – desks, chairs and shelving – then putting them back together, creating new relationships between these constituent elements in order to explore the possibilities afforded by future working spaces. Removing some of the shelves, for example, turns part of the shelving unit into a desk. Shelves and desks can mingle – and new, defined spaces appear between their overlaps. Some of the shelves may even become steps; sitting on them transforms them into benches; and when accessed from behind they form an enclosed desk.

Rain Bottle

Maison et Objet, Paris

A project for the Trend Exhibition space in the main hall of 2014's Maison et Objet, Paris. The exhibition theme, 'Words', asked participants to consider the relationship between language and design. Nendo chose the word 'rain', but considered its many nuances in Japanese – a language that has dozens of words for rain depending on the weather conditions and time of day – in order to create a design that reflected the fine subtleties and sensibilities of the language.

The exhibit consists of clear, acrylic bottles lined up, each exemplifying a different kind of 'rain' and distinguished by the inclusion of appropriate objects and patterns within them. *Kirisame*, *biu* and *kosame* all refer to varying degrees of fine drizzle, and are marked by tiny bubbles created by laser. *Niwaka-ame* is a sudden downpour; its bottle has thicker, vertical strands of glass. *Mizore* is sleet, and *yudachi* falls in the evening. *Kisame* is precipitation that drips from the ends of tree branches, tiny versions of which are included in the bottles, and *kaiu* is rain that falls mixed with dust and pollen – its bottle contains enlarged particles of these, floating among bubbles. Seasonal rains were also included – from the *samidare* that falls in the spring to *shigure*, rain specific to autumn and winter, and marked by tiny leaves.

By exhibiting twenty different kinds of 'rain', the intention was to express the depth of Japanese culture's profound relationship with nature.

419

Peg

Cappellini

An armchair with back legs like wooden stakes that pierce the sides of the backrest. These 'stakes' appear to hold the backrest's profile in a nipped-in curve that provides firm lower-back support and comfortable seating, despite its compact size. The Peg series was later extended to include a succession of sofas, beds, low tables, corner shelving units and mirrors. All the pieces are unified by their compact feel and legs like wooden tentpegs.

Stay-Brella

Moon Bat

An umbrella whose black, cylindrical handle splits at the end into two angled 'feet'. This feature ensures that it is able, when not in use, to stand firmly when turned on its handle. It can also hang securely from tables and stay propped up against a wall when furled. Lacking, as it does, the elbow 'crook' of traditional umbrellas, a handle-mounted wrist strap ensures that the rolled Stay-Brella also hangs easily on the arm. It comes in eight different colours, all with a surface coating that reduces UV rays by 90 per cent – thereby doubling as a parasol.

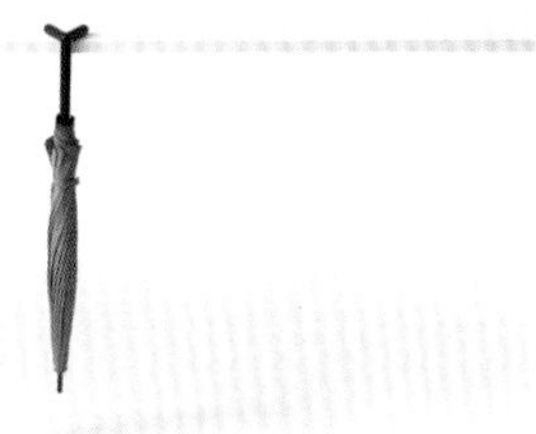

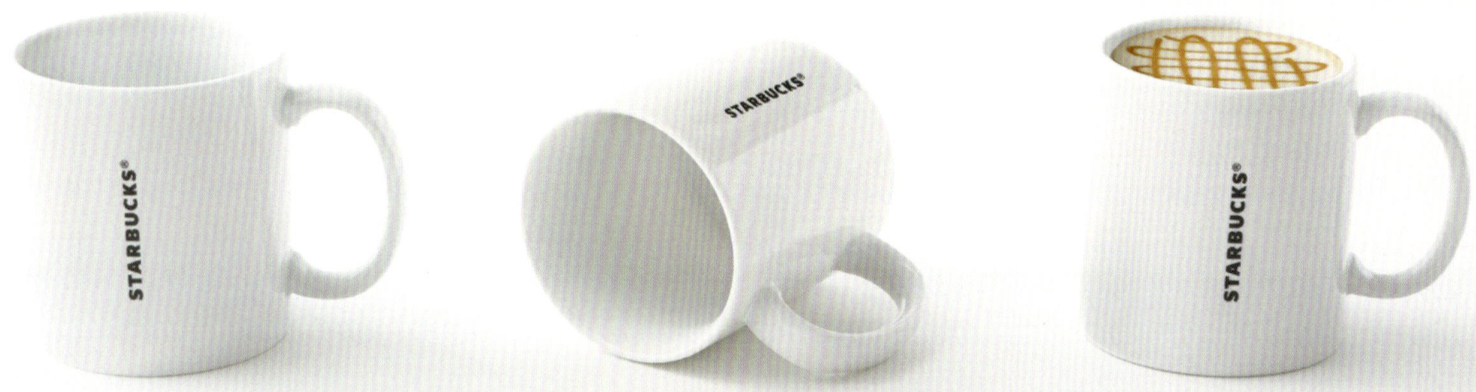

Mug Americano, Mug Latte, Mug Caramel Macchiato

Starbucks

An adaptation of the classic Starbucks' white coffee mug. By adding a graphic to the base that looks like the surface of a coffee drink, when the beaker is drying or being stored upside down, it appears as though it is actually full. To add to the illusion of the topsy-turvy effect, the Starbucks logo was arranged vertically, meaning it is legible regardless of which way up the mug is standing. The design comes in three varieties – Americano, latte and caramel macchiato – and was released in Starbucks cafes throughout Japan.

Hello Kitty Men

Sanrio

Hello Kitty is instantly recognizable, and one of the most famous characters in Japan. While many of her attributes naturally induce the feeling of *kawaii* (cuteness), this set of t-shirt designs aimed to appeal to a new audience by viewing her through a more rational, 'scientific' filter. Radiograph Kitty features an X-ray image, revealing her skeleton and shedding light on her mysterious internal makeup. Math Kitty uses numbers and symbols associated with math formulas to spell 'hello kitty'. Physics Kitty references Kitty's height as being equal to five apples, which are placed next to a scale. Evolution Kitty draws attention to Kitty's continuous development, superimposing her on the classic illustration depicting human evolution.

Contrast Ruler

Mark's

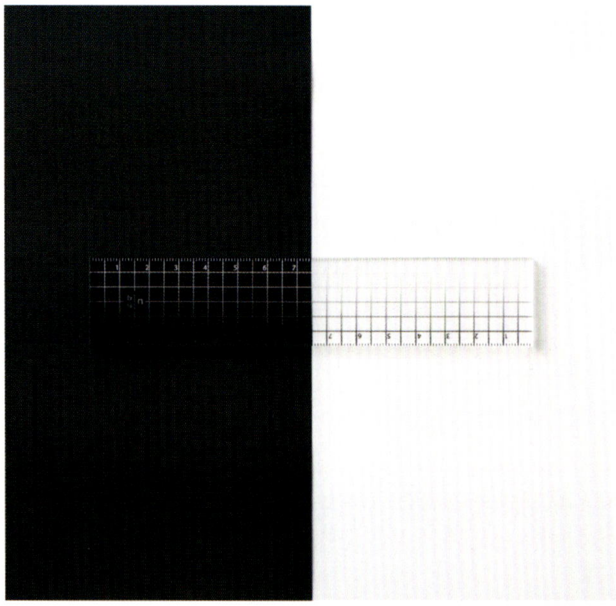

A clear acrylic ruler with measurements that gradually change from white on one side to black on the other. This allows them to be visible, and the ruler correspondingly easy to use, on both dark and light surfaces.

Link Clip

Mark's

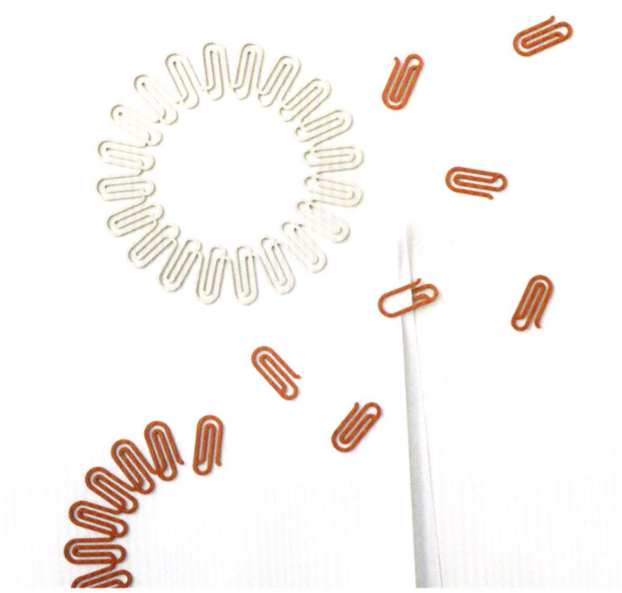

A set of twenty paperclips that together form a continuous circular pattern, from which they can be detached one by one as needed. They are actually made of paper, making the clips easy to tear off, as well as recyclable. To ensure that they work as well as metal clips do, a high-friction paper was selected as the material.

Outline Tray

Mark's

Despite using them every day, we don't usually pick up desk tidies full of pens and paperclips and walk around with them – so why do they need bottoms at all? Based on the realization that they are really just for separating out things by type, the Outline Tray was designed as a frame for filing small objects, rather than a container. The frame form is easier to clean than other organizers, and it is less visually obtrusive on a work surface. When not in use, the Outline Trays can be stacked inside one another for compact storage.

Cubic Rubber-Band

Mark's

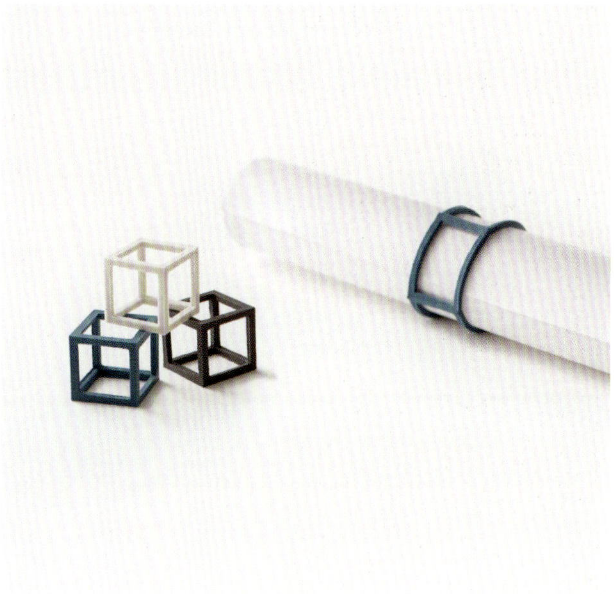

An assertively three-dimensional, cube-shaped rubber band. Its geometrical shape makes it easier to find in a messy drawer, and much simpler to pick up from a surface, than its limper, more conventional counterparts.

°03
Dark chocolate / Rhubarb / Violet.
¥1,300
°08
Milk chocolate / Speculoos / Orange
¥1,300
°15
°25
°19
°05
Dark chocolate / Cherries / Lemon.
¥1,300
°09
Milk chocolate / Passion fruit
¥1,300
°23
°14
Dark chocolate / Cherries / Cinnamon

BbyB

BbyB

The first overseas outlet for BbyB, the chocolate shop of Belguim-based, two-Michelin-star chef Bart Desmidt. With their identical bar form and standard range of colour packaging, BbyB's chocolates may look uniform from the outside, but they come in thirty distinctive flavours. Being all the same shape, their packaging is modular: five bars of chocolate slide neatly into each box, and five boxes slot together into a cube. The contents become apparent only gradually, as the boxes are opened and closed, offering surprises and turning the cube into a 'magic chest of drawers'.

Accordingly, the shop itself was turned into a three-dimensional version of the packaging. The confectionery seems to float in a transparent 'chest of drawers' at the centre of the store. When making a purchase, customers slide out the chest's drawers and remove the chocolate themselves. Towards the rear, this transparent chest is replaced by a single, mid-height display surface, which then turns once more into a countertop in the shop's cafe space.

The front of the premises is entirely white and the cafe at the rear entirely black, following the colour scheme of the cubes' outer packaging. The white display wall is 'tiled' with chocolate packages, so that its entire surface turns visually into yet another 'chest of drawers'. The design creates a seamless transition between packaging, purchasing and partaking – the individual spaces following on one from the other in a compelling consumer experience.

2015

Chocolatexture

Maison et Objet, Paris

Many factors determine a chocolate's flavour – the cocoa's country of origin, its percentage, the techniques used by the chocolatier. The focus of this project was not on such common factors, however, but on the effect of a chocolate's shape on taste. The nine different types included in the set had the same overall dimension – 26-millimetres (1-inch) -cubed – but featured pointed tips, hollow interiors, and smooth or rough surface textures. Indeed, each was named after a Japanese expression used to describe texture – for example, 'sube-sube', smooth; 'zara-zara', rough; 'toge-toge', sharp; 'fuwa-fuwa', light/airy; and 'suka-suka', look through/hollow.

Launched at the 2015 Masion et Objet Designer of the Year awards ceremony, the experimental chocolates were presented as part of a pop-up Chocolatexture Lounge. The seating area was surrounded by a 'forest' of 2,000 8-mm (5/16-inch) -thick aluminium pipes of varying heights, each carefully painted in chocolate-coloured gradations and arranged to create the feeling of a large, rippling, molten-chocolate wave. It was neither a cafe, an exhibition nor an installation, but a place in which to experience and enjoy the product through as many senses as possible.

Tubu-tubu
Sube-sube
Zara-zara
Toge-toge
Goro-goro
Fuwa-fuwa
Poki-poki
Suka-suka
Zaku-zaku

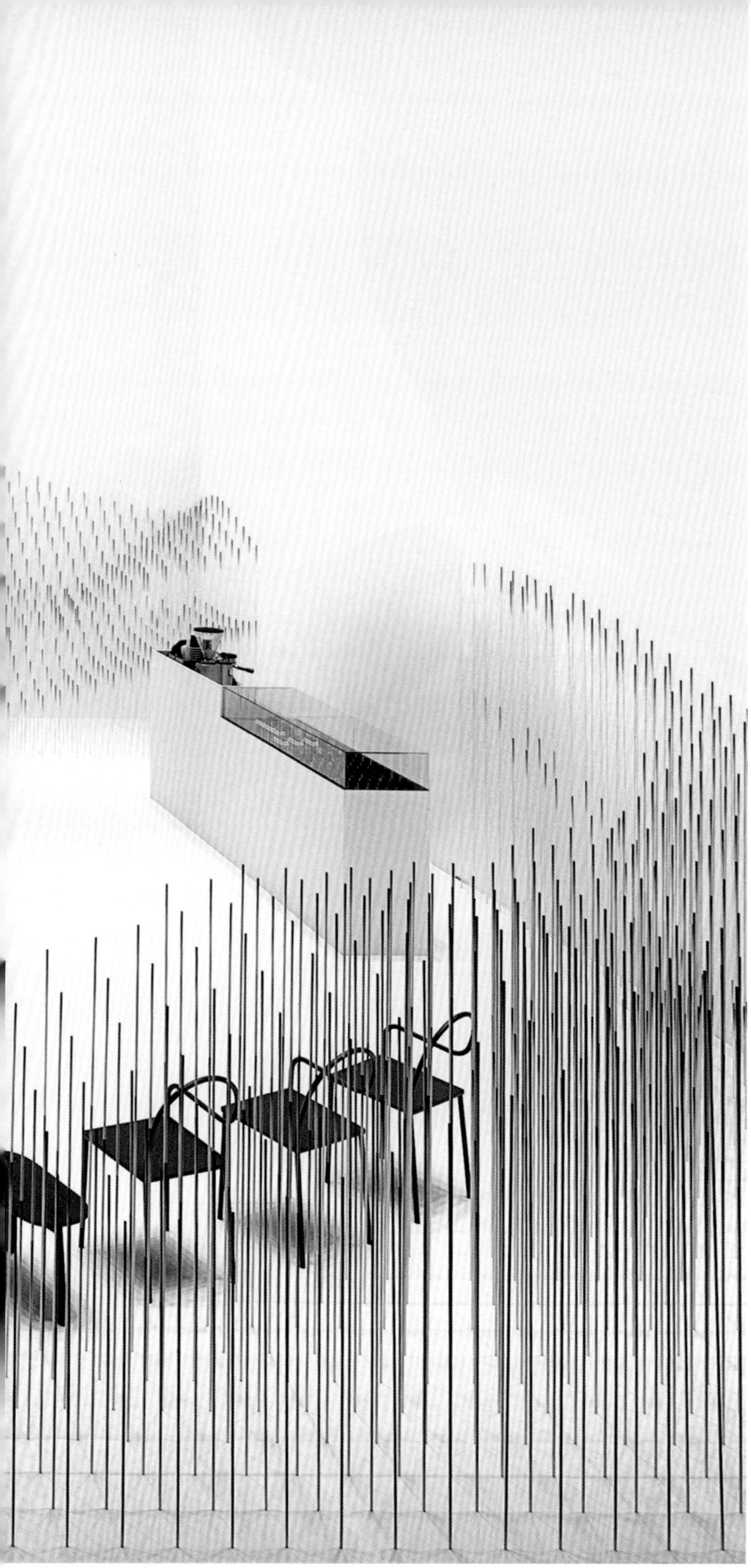

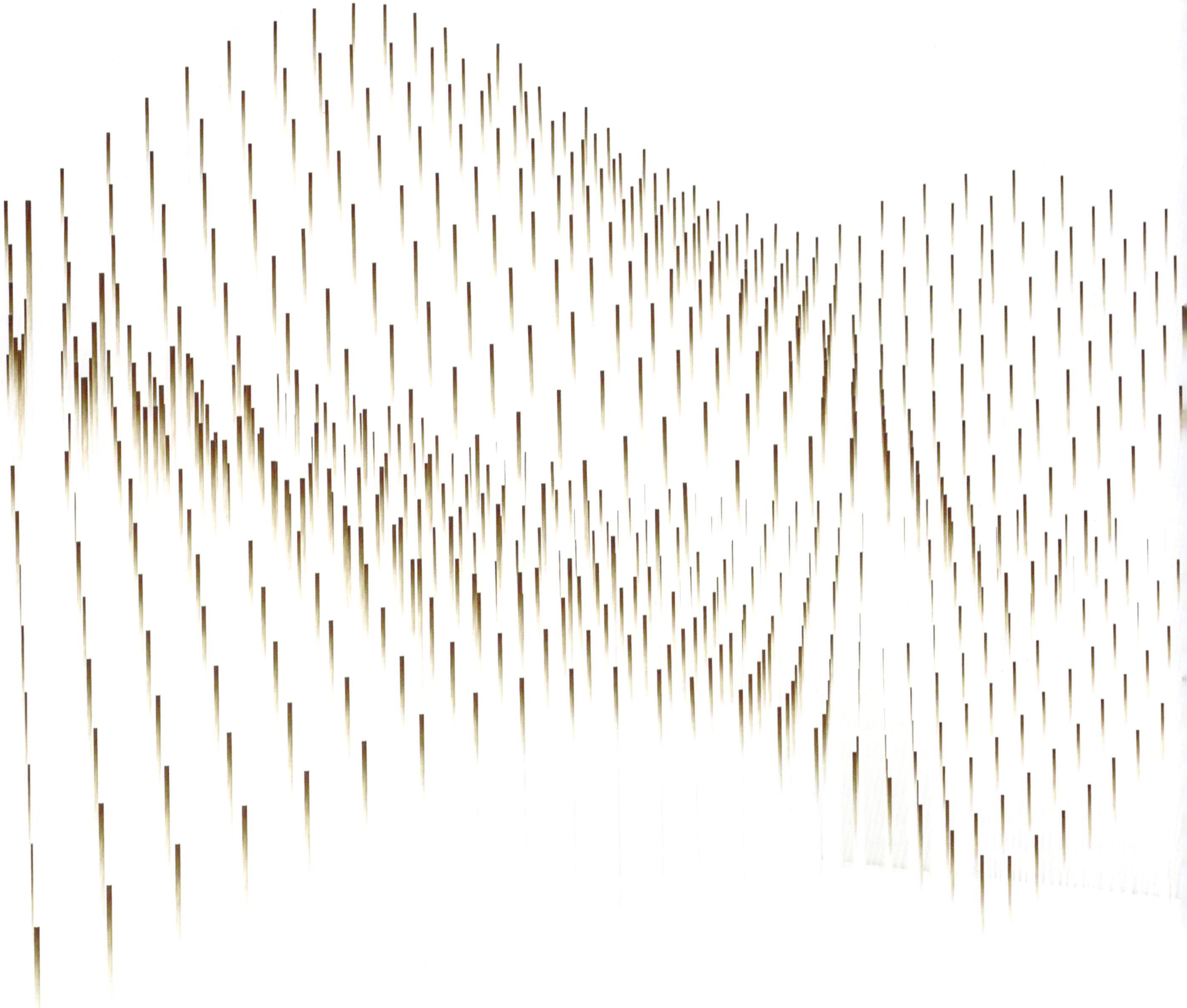

Single-Curve

Gebrüder Thonet Vienna

A collection – including a stool, bar stool, coffee tables, and a dining table – designed for the Austrian furniture brand Gebrüder Thonet Vienna, well-known for its bentwood products. While the brand's trademark technique remained, only a single element of each piece was bent, rather than the overall structure. The combination of straight lines with a single bent section created a strong sense of contrast, which – along with close adherence to the materials and proportions of the classic Thonet design – further strengthened the character of this ensemble.

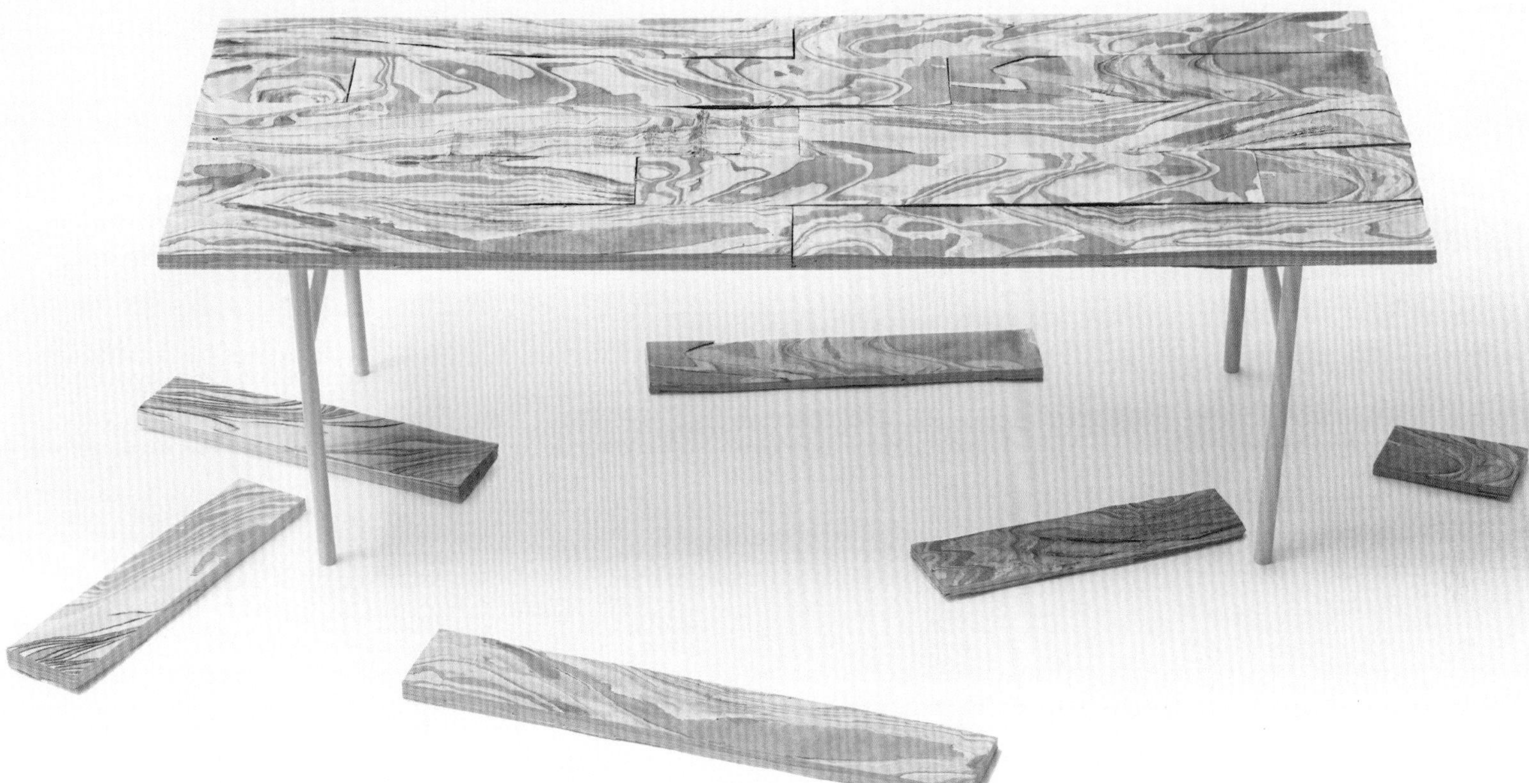

Alcantara-Wood

Alcantara

A bench design premiered at the exhibition *Alcantara, Technology of Dreams*, held at Milan's Palazzo Reale as part of Expo Milano 2015. Alcantara is a microfibre produced by the Italian manufacturer from which it derives its name. The aim of the design was to demonstrate the material's functional values, which include a greater resistance to fraying at the edges than ordinary fabric. Sheets in blue and grey tones were layered and rolled to create a 'log', then sliced to create 'Alcantara timber'. These slices were then assembled to create the bench's surface, showcasing Alcantara's uniquely versatile qualities.

Corners

Moroso

Placing an item of furniture in the corner of a room will sometimes help to balance the space. Likewise, there is often a certain surface that just feels right for placing small, daily-use items such as a book, mobile phone, tablet, set of keys or an alarm clock. A space surrounded by three sides often provides more security compared with an exposed, flat surface such as a table or tray.

Corners is a piece of furniture that resembles just such a 'corner of space'. Its three-sided cube box is designed so that it can be augmented in various directions, allowing the box to be configured to suit the user's preferences and needs. The diminutive size encourages the use of several units, arranged freely or lined up next to a bed or a sofa.

NJP
Louis Poulsen

An aluminium desk lamp with the ability to emit both direct and diffused light simultaneously, inspired by Arne Jacobsen's iconic AJ Table lamp. It was designed for Louis Poulsen, the Danish lighting manufacturer founded in the nineteenth century, that originally commissioned the classic Jacobsen design, and which wanted to update it for the modern workspace.

The NJP's task-lighting shade has a 'cowl' at the top, which allows any heat to evaporate while also softly dispersing the ambient lighting. Unlike on other modern lamp designs, the power switch is a sturdy, clickable button with three simple modes – 'Off', 'Dim' and 'Bright' – making it easy to use. The button is placed subtly, where a thumb rests when a user's hand holds the shade to adjust its height.

Tokyo Tribal
Industry+

A furniture collection for Singaporean design company Industry+, which used solid oak for the main frames and volcanic-sand plaster for the surface finishes, in combination with bamboo rattan hand woven by local artisans in the Philippines. The bamboo's elastic properties make it ideal for such things as lumbar support in chairs. It was also used as a design feature – in one item, for instance, the table legs pass through a bamboo basket attachment. Through such details, these pieces seemed to marry the twin concepts of furniture and interior objects, whether practical or ornamental, with the rattan baskets adding functionality to tables and stools. In allowing these various products and materials to converge and function together, the aim was to create a sense of a small and tightly-knit 'tribe' of objects.

Fragment
Glas Italia

A screen made up of two layers of transparent glass. In the first layer, a pattern consisting of small rectangular mirrors was vapour deposited on the glass – a process (Physical Vapour Deposition – PVD) in which solid or liquid material is vaporized in the form of atoms or molecules, and transported as a vapour through a vacuum onto a substrate. For the other layer, a similar pattern was deposited but rotated through 90 degrees.

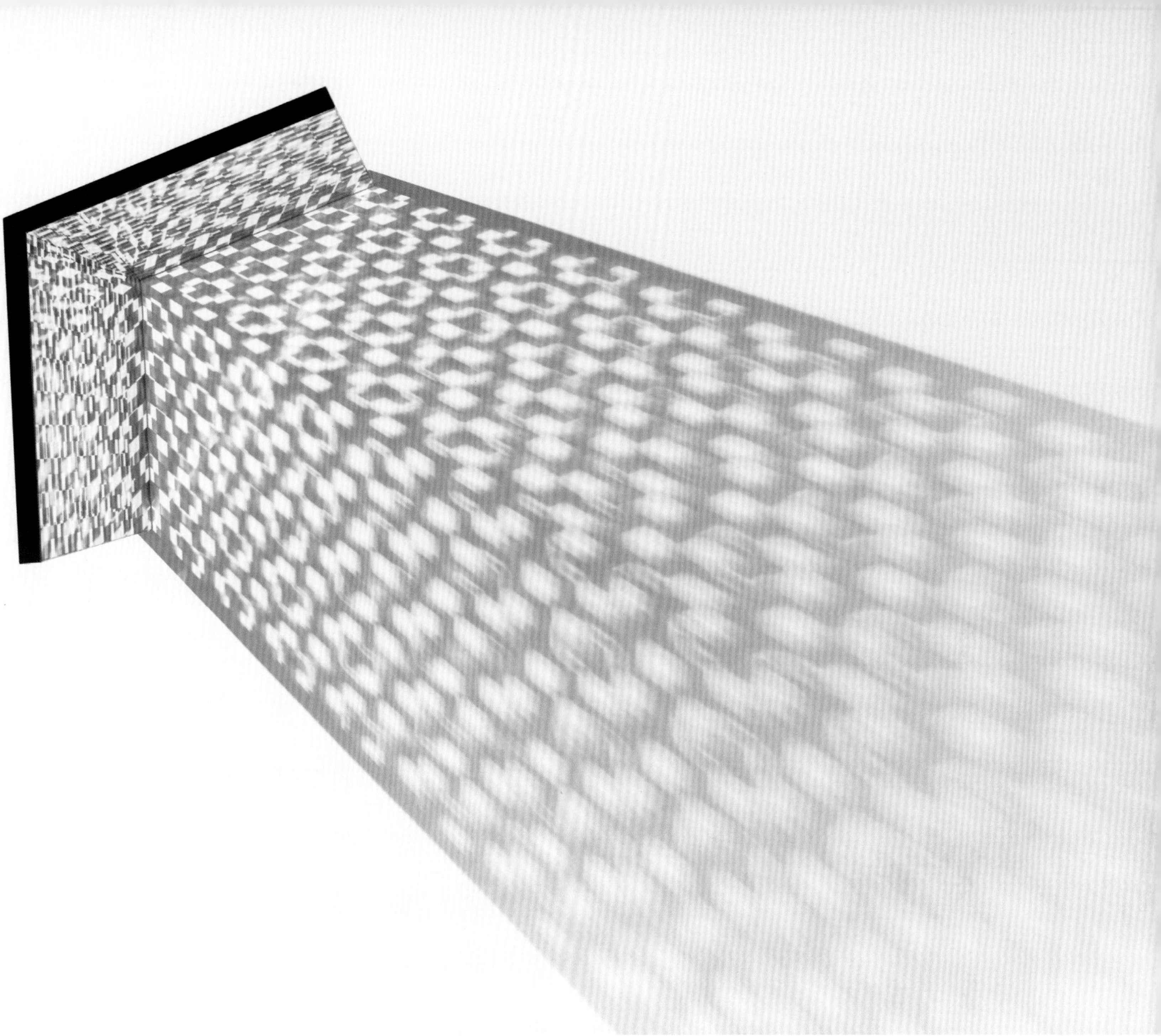

However, these were no ordinary PVD processes: normally, when a mirror is vapour deposited onto glass only its surface becomes reflective and the reverse side turns grey. By using a special PVD method that allows both sides to become mirrors, Fragment's reflective properties can be enjoyed from any aspect. As a result, three different effects exist on the screen's surface: one that allows a clear line of sight through both layers of glass, one that reflects conventionally and one in which the reflection itself is reflected. When the viewer stands in front of the partition, their own image and that of objects beyond the screen reflect in a fragmented manner, changing their appearance depending on the angle at which they are viewed.

Slide

Glas Italia

A collection of black-glass shelves and tables, each featuring one element that has 'slid' out of place. However, a corresponding, transparent piece – shelf or tabletop – covers the exposed area, so that items can still be placed on it. The shelves were particularly challenging to develop, since the displaced black-glass element must be attached to its transparent-glass support along an extremely thin edge.

Layers

Glas Italia

A series of cabinets made entirely from glass, with multiple mini-screens hanging at the front, middle and rear. Since the tones of each coloured-glass screen differ subtly, and they slide left or right, a wide range of hues emerges depending on how they overlap. This colouring technique, which involves inserting tinted films between two sheets of glass and heating them, is a speciality of the manufacturer, Glas Italia.

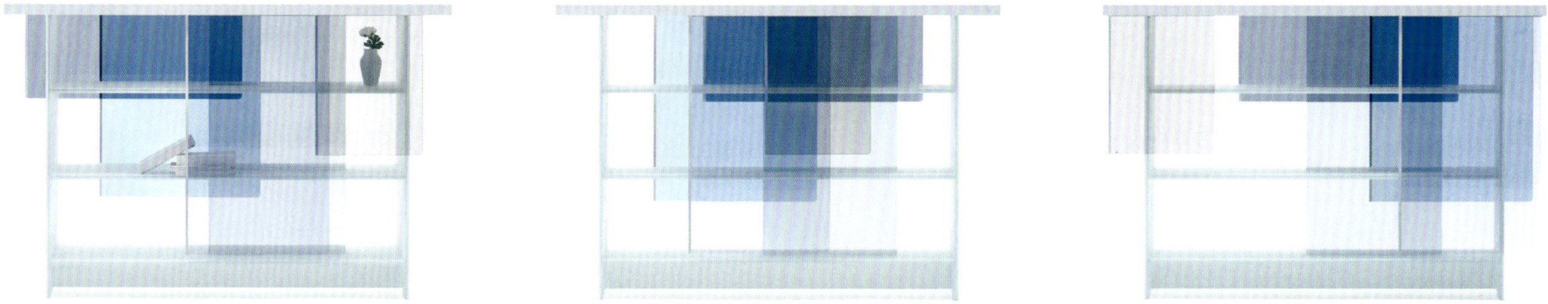

Seven Doors

Abe Kogyo

Seven door designs commemorating the seventieth anniversary of Abe Kogyo, a manufacturer of wooden doors. Starting from a basic default model, each one was customized according to the various techniques that the company specializes in.

'Lamp' is a combined door and lighting fixture, employing the kind of wiring techniques used in electronic locks to enable the light to function while attached to a hinged, free-swinging object.

'Baby' allows adults and children to walk through doors that match their respective sizes. It references Abe Kogyo's sideline of nursery and pre-school interior fittings.

'Slide' works like a window blind, creating small spaces to let light or air in, and to create a greater sense of connectivity between rooms.

'Hang' is lined with a thin, internal magnetic sheet, allowing the user to attach accessories to it, such as trays, dustbins, flowerpots and vases.

'Kumiko' is named after the delicate technique of assembling grooved and mortised wooden lattice pieces without nails. The resulting pattern gradually blends into the plain portion of the door.

'Wall' covers the door with wall fixtures such as shelves and picture frames, thus blurring the very concept of a door.

'Corner' allows the user to enter and exit through a room's corners. As it opens particularly widely, an additional practical benefit is easier and more comfortable wheelchair access.

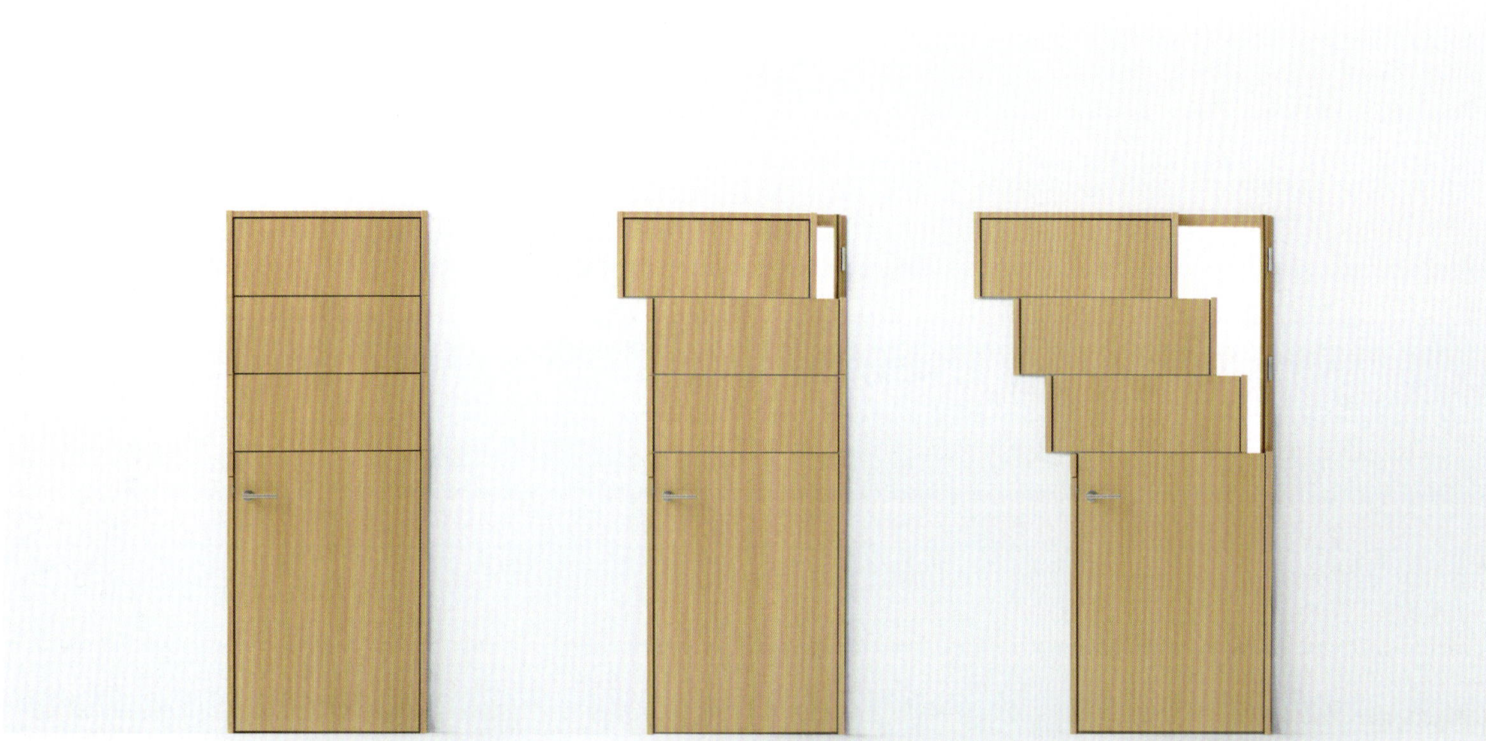

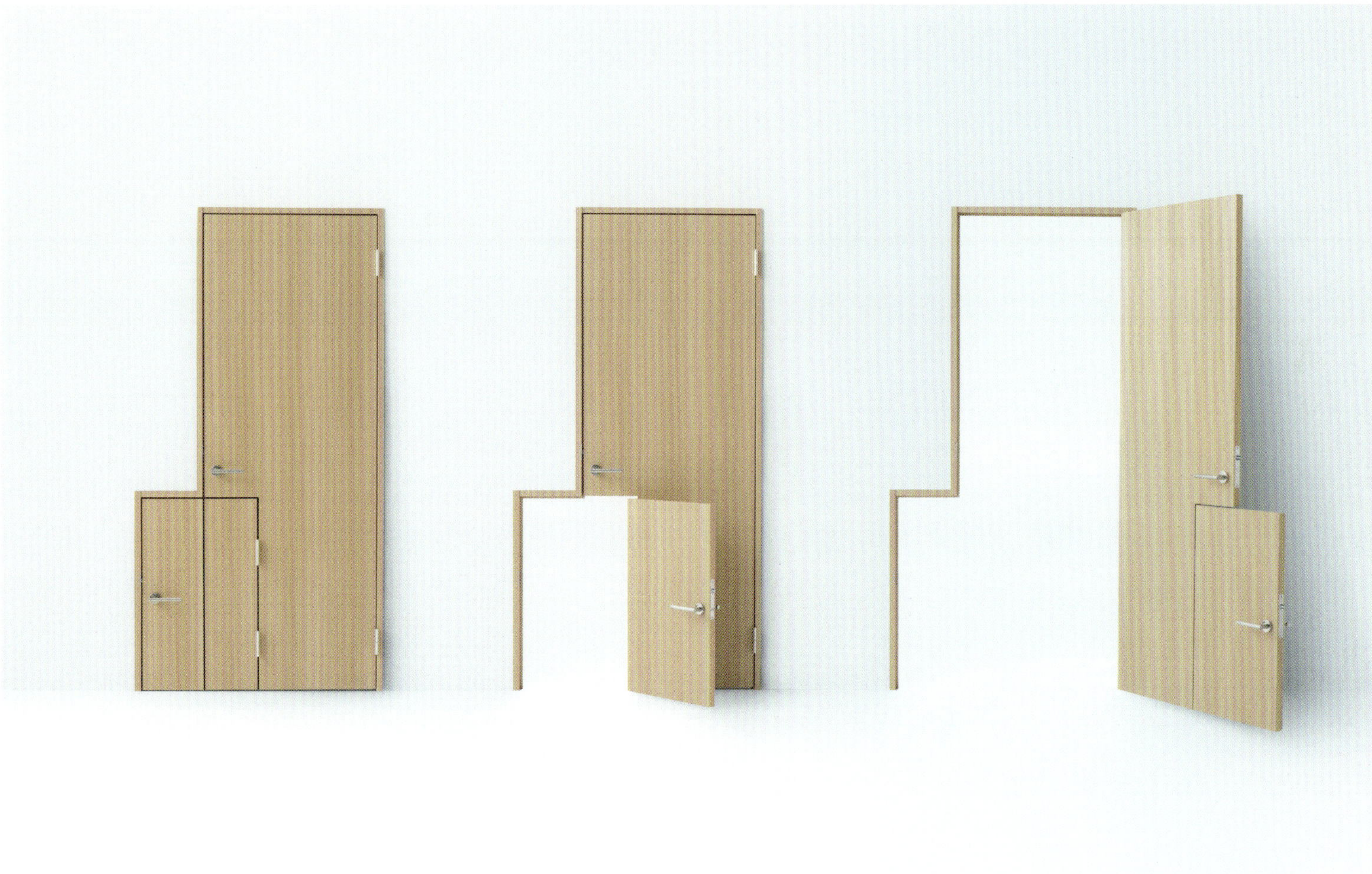

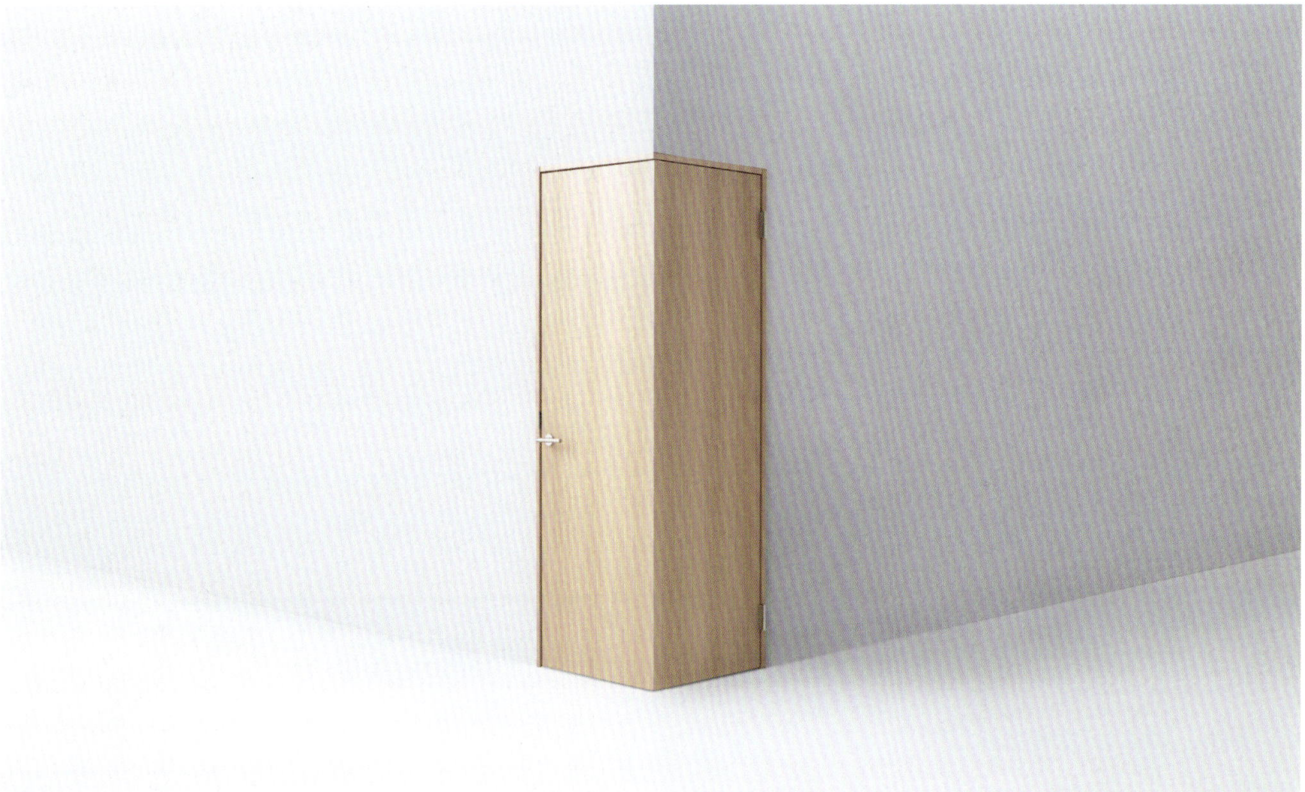

Sticker-Calendar

The Pack

A six-part calendar with small, daily peel-away stickers. Peeling off one sticker at a time gradually reveals the scenery of the following month decorating the page underneath, allowing users to feel the gradual changing of the seasons as each fades into the next. Removing stickers also uncovers special illustrations on certain occasions like Valentine's Day or Christmas, and a whimsical cat can be found hidden somewhere on each page. Twists like these are interspersed throughout, adding a further element of fun.

Sunset

Grasse Tokyo

Sunset appears to be a normal white candle. But once lit, its centre begins to change colour as it burns – first to yellow, then orange, red, purple and, finally, to blue. These successive hues reflect faintly off the white wax that surrounds the candle. This transition of colours is inspired by the shifting shades of light that paint a sky at sunset. Each has its own accompanying scent: bergamot, lemongrass, sweet marjoram, lavender and geranium. Not only does the candle provide illumination, therefore, but the succession of colours also serves as a reminder of the passage of time.

Suntan-Pattern
Tanners' Council of Japan

One of the defining qualities of natural leather is that the change it undergoes with years of use does not equate to a reduction in quality but rather enhances its character, and this wearing-in actually results in a more attractive piece of material. To exploit this characteristic, the leather items in this collection – among them folders, containers and footstools – were painted with tanning oil and sunscreen in order to create an invisible pattern on the surface. At the time of purchase, the products appear to be plain leather but, with use and exposure to sunlight, the pattern gradually begins to appear. As more time passes and with further exposure, it disappears again due to overall fading.

Colourful Shadows

Expo Milano

The Japan Pavilion gallery space and a sixteen-product collection, designed for the Expo Milano 2015. Following that year's theme, 'Feeding the Planet, Energy for Life', products – including tableware and kitchen utensils – were centred around the concept of food. All the items exhibited were produced in collaboration with a different traditional Japanese craftsperson or industry and demonstrated the characteristically painstaking processes and techniques of these traditions in their material and textural finishes. To best convey their attention to detail, the items were coloured black in a reference to the novelist Junichiro Tanizaki's

1993 essay 'In Praise of Shadows', which includes a passage about eating a *yokan* (a sweet made from black-bean paste) in the dark in order to develop a keener palate.

Included in the collection were the curved CS002 bowl, designed to bring out the distinctive gloss and sheen of traditional Japanese lacquer; the CS006 cutlery set, made from aluminium treated with a urethane baking finish, which accentuated the texture resulting from this process; CS007's small, disposable plates or ornamental trays, dyed with traditional black calligraphy ink in order to soften the paper material, and then compressed to form the shapes; CS012's glassware vessels that focused on the aesthetic quality of a single cut: they comprised simple, two-layered cylinders of transparent and black glass, etched with one sweeping stroke; and CS015's matching teapot and cups, made using traditional metalware techniques, but with metal handles, which stretched out from inside the containers.

The *Colourful Shadows* of the exhibition's name signified shedding light on the intricacies of Japanese craftsmanship that often remain 'hidden' in the dark from the outside world.

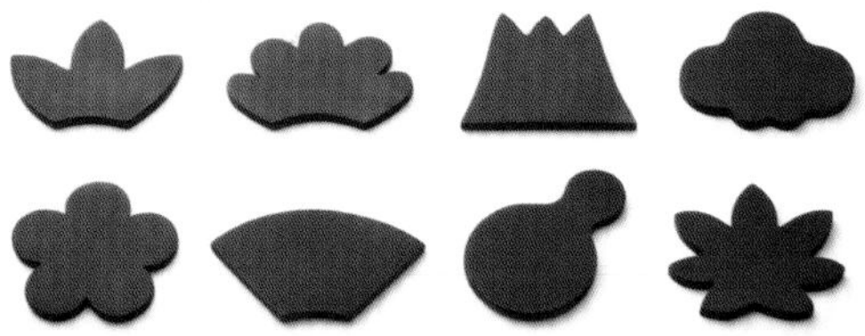

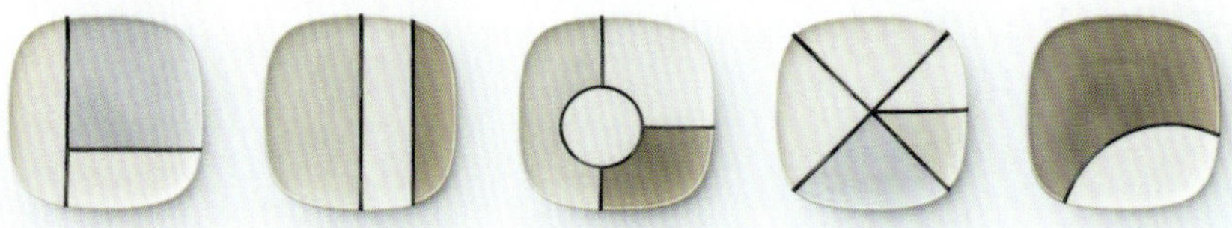

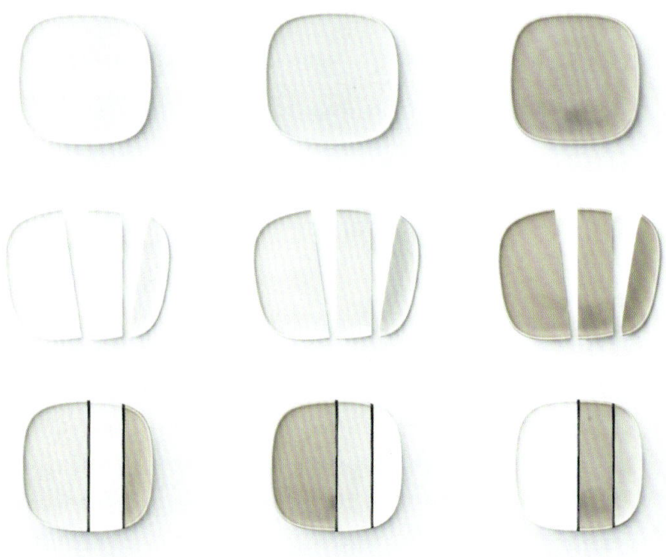

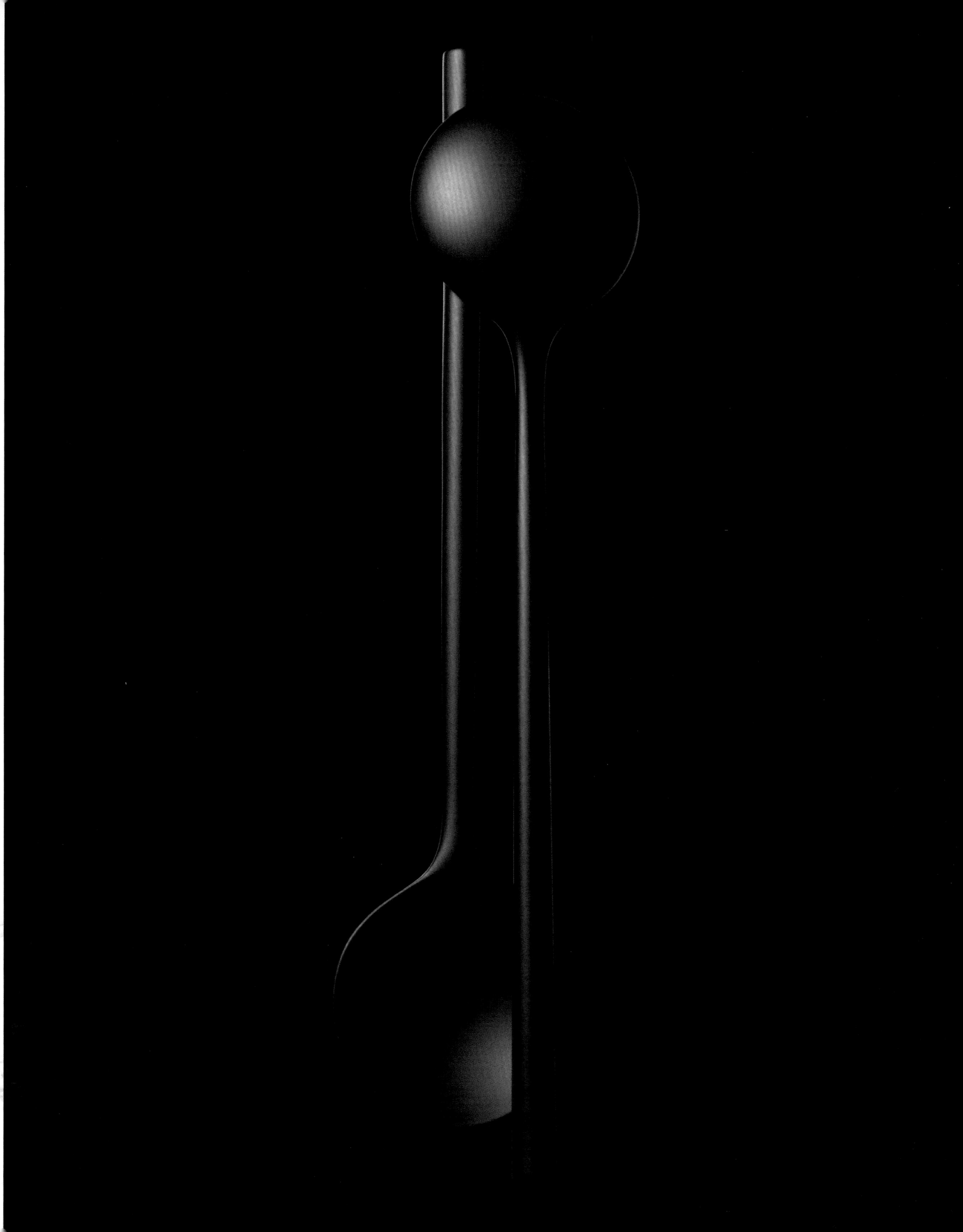

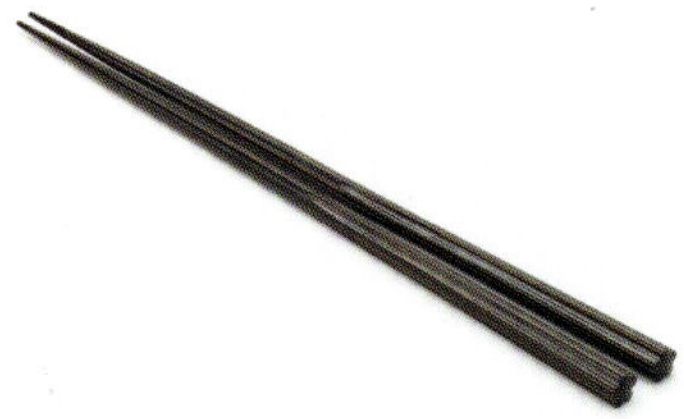

468

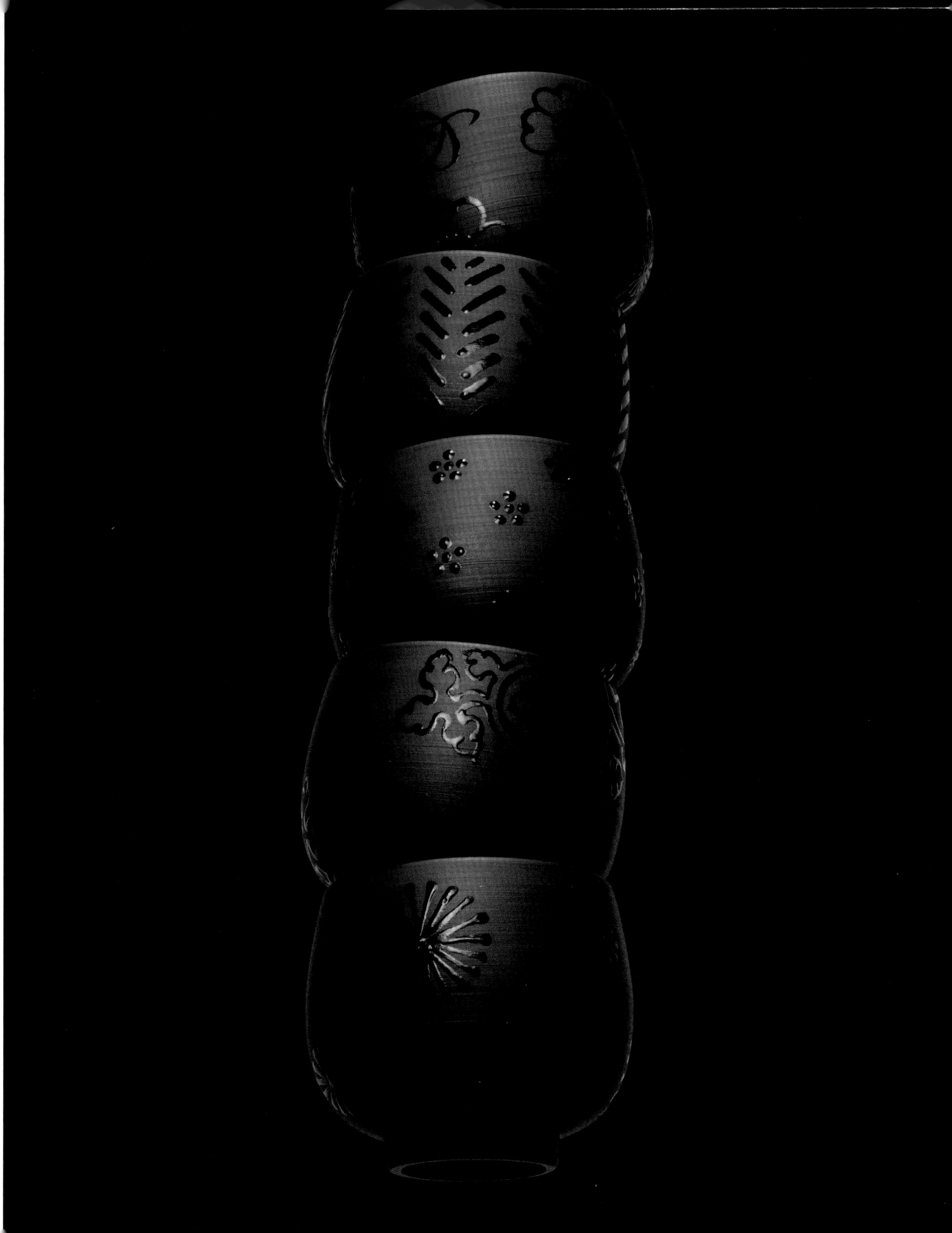

Offset Frame Chair

Kokuyo

A simple stacking chair with multiple uses. Although its composition is straightforward – comprising a plastic, resin shell with a one-piece back-and-seat section and a metal structure – its frame is set slightly 'protruding' beyond the shell in all directions. This gap allows the chair to be easily gripped for stacking, unstacking or carrying around. It even facilitates hanging objects like blankets or garments on the chair. The collection comprises eight different models, ranging from armchairs to castor-mounted office seating, with six colour variations for the shells and three for the frames – as well as detachable fabric and leather covers.

Float

Moroso

A seemingly plain stool composed of a rectangular, plywood seat and steel-pipe legs. Two legs have been cut off at the top, directly underneath the seating surface, leaving a gap that suggests a 'floating' seat when viewed front-on. The cantilever afforded by the continuous rear legs, with its bouncy quality, allows for a more comfortable seating experience than that offered by a conventional stool.

Kame

Fabbrica Pelletterie Milano

A suitcase appropriately named Kame ('tortoise' in Japanese). Made of polycarbonate and textile, its shell is light yet strong and the cloth covering is flexible for easier handling than a normal, fully rigid case. Due to the soft foldable cover, in fact, this case can be opened from both its top and side, allowing easy access to its contents when upright, even in tight or crowded spaces. A cable-lock mechanism can be used, in conjunction with Kame's ring zipper, to secure the suitcase for travel; it may also be deployed to tie it to a post or railing if the user is required to leave it out of sight for a few minutes.

By|n Cosmetics

Kozuchi

A body-care product line that uses plant-based, natural ingredients and essential oils. Judging that extravagant packaging would not suit the concept, the design instead adopted a simple and refined look. Each container features a label cut out to form the first letter of each product: S = shampoo, C = conditioner, B = body soap, L = body lotion, H = hand soap and h = hand cream.

Carrera Time Machine

Tag Heuer

A watch design based on the Carrera, one of Tag Heuer's boldest and sportiest models. The face, case and wristband remain completely unchanged, however, the watch hands have taken on an elaborate, baroque form – as if an elderly, traditional watchmaker has repaired the timepiece and put on a pair of hands from a vintage model by mistake. To accentuate this sense of incongruity and strangeness, the Carrera Time Machine's packaging references an old-fashioned, sleeved book or a glass clock stand, individually handcrafted from an antique jar.

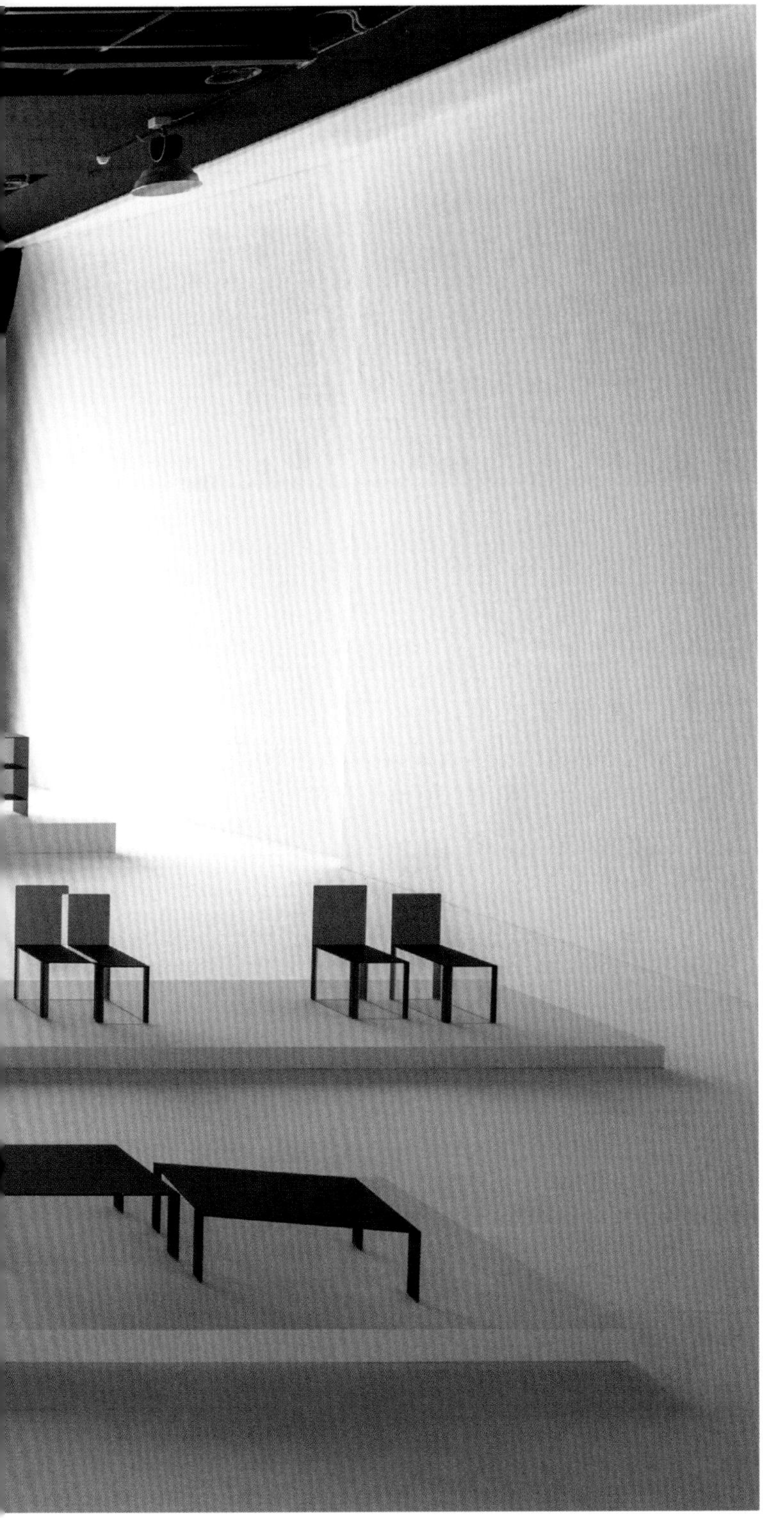

Pair

Glas Italia

A collection – including items such as chairs, tables and shelves – designed around the idea of two similar pieces of furniture that intersect one another. Made primarily of frosted and smoked glass, transparent glass was used for the overlapping sections. The different types of glass are attached using only their edges, demonstrating outstanding technical manipulation of cut and adhesion.

476

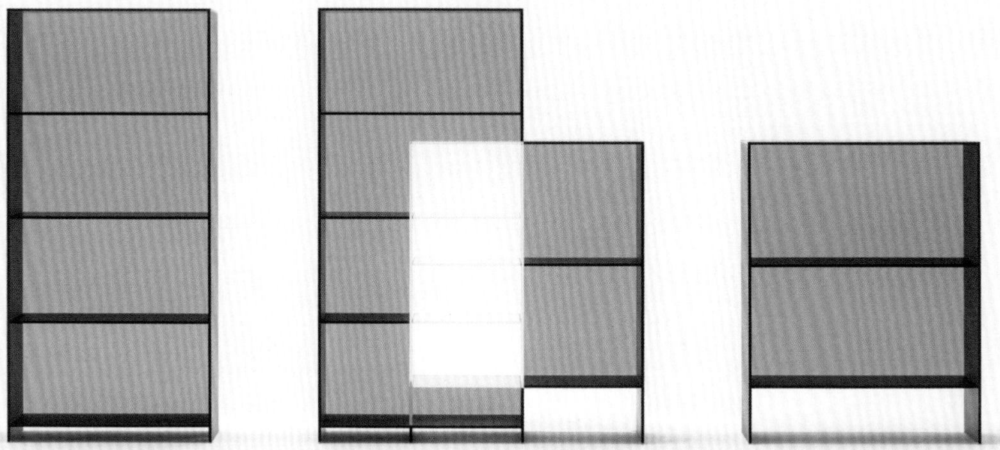

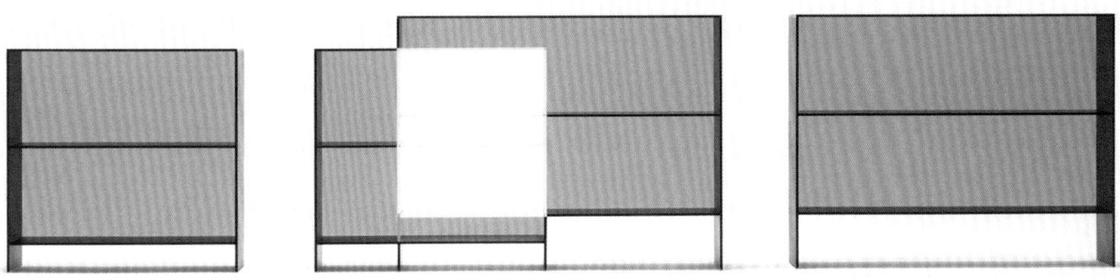

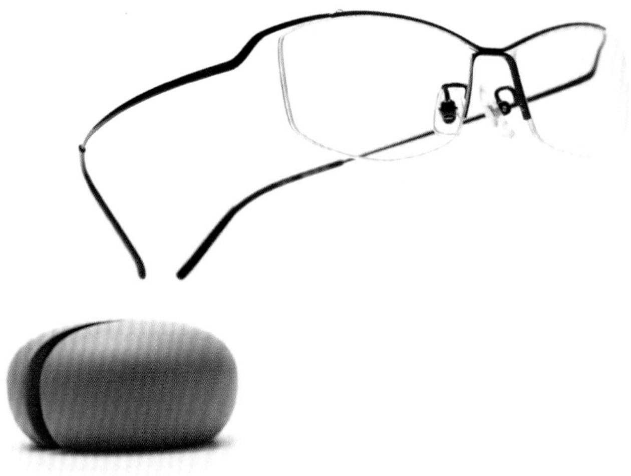

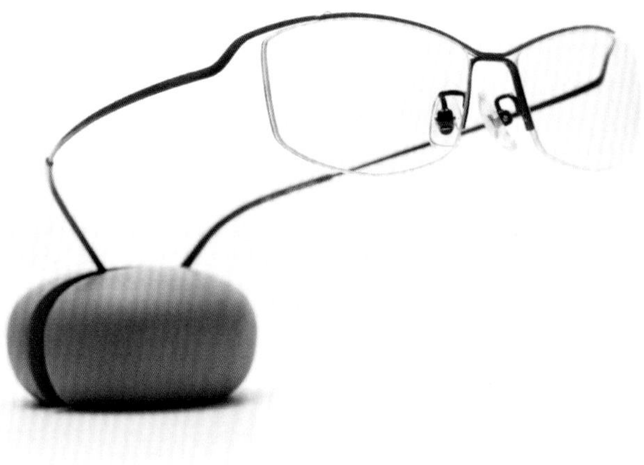

Zerogra-Stand

Meganetop

An optical frame stand for Zerogra eyeglasses that utilizes the lightness and resilience of their titanium frame and single-body construction. The result is a small, rounded form giving the impression of a pebble or drop of water. By placing the tips of the frame into the central slot, the glasses are fixed in position as though floating. The stand's size was minimized such that even if multiple stands were lined up in a row, with each of them holding spectacles, they would not take up too much space – and, when not in use, would not be obtrusive.

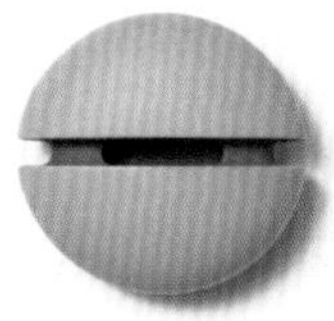

Zerogra-Case
Meganetop

A ring-shaped glasses case, devised in response to the tendency of the Zerogra glasses' elastic frame structure to push outwards. When the spectacles are placed in the wide hole of the case, the frame's outward push ensures that they remain firmly in position. Compared with the previous Zerogra case, the redesigned version reduces bulkiness – and the glasses can be accessed without the need for opening and closing a lid.

Minim+Aid

Sugita Ace

Japan has always been faced with natural disasters. On 11 March 2011, the Great East Japan Earthquake once again reminded us of the importance of disaster and emergency preparedness. However, rather than conventional emergency supply kits, which tend to resemble one another, people are now seeking a more versatile solution, appropriate to many situations. This necessitates the development of a kit that includes the bare minimum for a city dweller to reach a place of refuge during an earthquake or other disaster.

The Minim+Aid collection responds to that need, comprising a whistle, radio, raincoat, lantern, drinking water and plastic case – all packaged inside a 5-centimetre (2-inch) -diameter tube that is waterproof and floats. The radio has manual charging functionality, which can also be used to power smartphones, lanterns or other devices via USB. The case can store medicine or other necessities, and the tube in which the drinking-water foil pouch is stored can also serve as a cup. The kit is easily portable, can be worn over the shoulder using the strap provided and is easy to keep near a building entrance, ready to go at all times – just leave it in the umbrella stand or hang it on a coat rack.

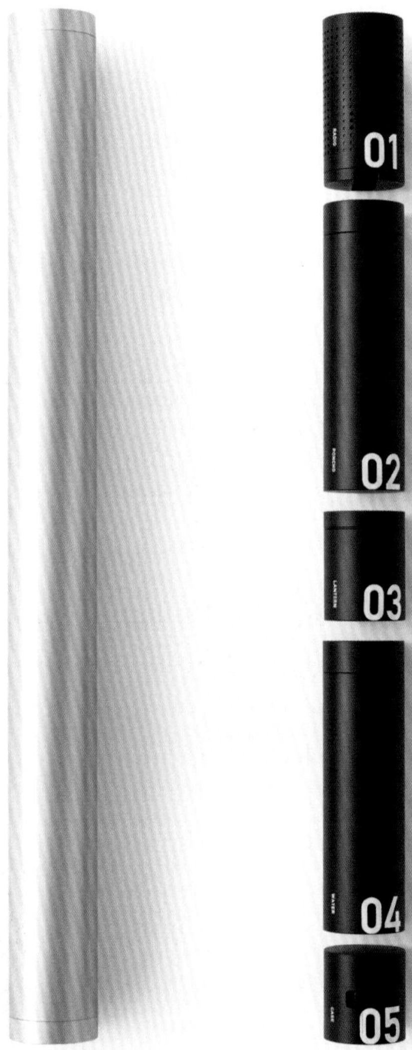

01

RADIO

02

03

03

04

04

05
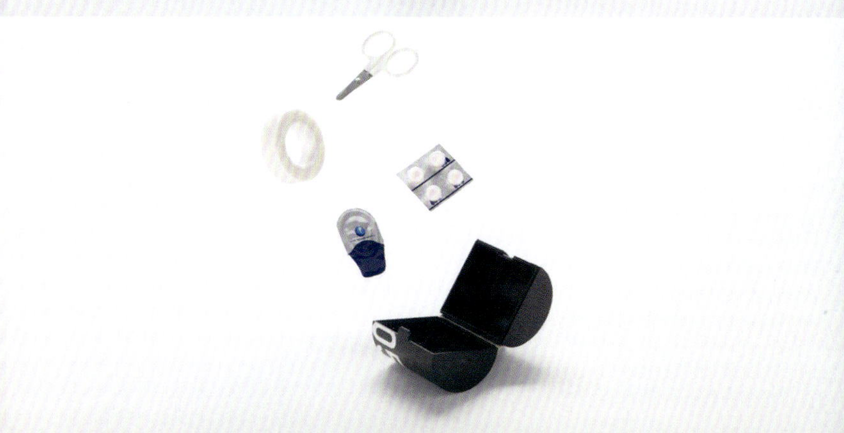

Triangle-Roomshoes

Senko

Cone-shaped indoor footwear that looks triangular from side-on. The shape allows the shoes to be slipped on easily, and forms a natural crease in the fabric when worn. The 'roomshoe' can stand on its own without taking up too much space, and several can be stacked up in the hallway of a residence with no need for space-hungry slipper racks. Materials were selected for their comfort and steadiness when stood on their own.

Shiawase Banana

Unifrutti

Flame

Grasse Tokyo

A packaging design for the high-quality, award-winning bananas that grow on Mindanao Island in the Philippines. As environmentally friendly bananas – using only organic fertilizers, and with a strict restriction on pesticides – they became the first such fruit to be awarded an impressive two stars at the International Taste and Quality Institute (iTQi) Superior Taste Awards in Belgium.

The packaging has, therefore, been reduced to a minimum and applied as a sticker on the surface of the fruit, imitating its exterior texture and even occasional bruises. When peeled back, printed information about the banana is revealed, compelling the consumer to read the story behind it. The healthy fruit combined with the playful packaging inspired the use of the Japanese word *shiawase* (happiness).

A ceramic scented-candle holder with a flame-shaped opening on one side. This detail makes it easy to light the candle once it is inside the holder but also increases airflow, helping to diffuse the candle's scent.

Nest Shelf

London Design Festival

A extendable shelving unit that can increase in size anywhere up to double its original width. Its vertical supports are made from 3.5-millimetre (around 1/8-inch) -thick carbon fibre, while the 3.7-millimetre (1/8-inch) -thick horizontal elements consist of aramid-fibre-made honeycomb material sandwiched between two layers of carbon fibre. Both sections are coated with a larch veneer, resulting in 4.7-millimetre (3/16-inch) -thick panels partly hidden within the outer layer. The Nest Shelf allows the user the flexibility to choose a width appropriate for the space available. Extending it halfway out produces a grid-style layout with rows of square shelving compartments.

Twig
Alias

Since the Italian furniture brand Alias specializes in products made from aluminium, the idea was to create a chair made of four aluminium 'twigs' and a seat. While the metal structure of all the chairs is the same, five variations of its upper-portion units feature differing shapes, colours and materials. A number of diverse constructions and sitting experiences are achieved by fitting these units on to the four legs in various ways – creating, for example, individual T-shaped back-rail elements 'growing out' of the tops of the legs; continuous, single-level back-rails; or broader backrests, both high and low. Choices for seat material include wood, plastic and fabric. When the seat and upper portions are removed, the aluminium frames of multiple seats can be stacked – making storage convenient and reducing potential transportation costs.

Soft

Glas Italia

A series of box-shaped low tables using sheets of frosted glass. At the joints between two sheets, their bonded edges are cut at an angle of 45 degrees and printed with bright colours. These colours have a lengthwise gradation effect – such as from purple to red, orange to yellow and blue to purple. Additionally, the reverse of the frosted glass is printed with a gradated colour that strengthens at the edge to accentuate the effect of these colours 'bleeding' out from the joints. The combination of these two techniques contradicts the conventional image of glass as a hard, sharp material.

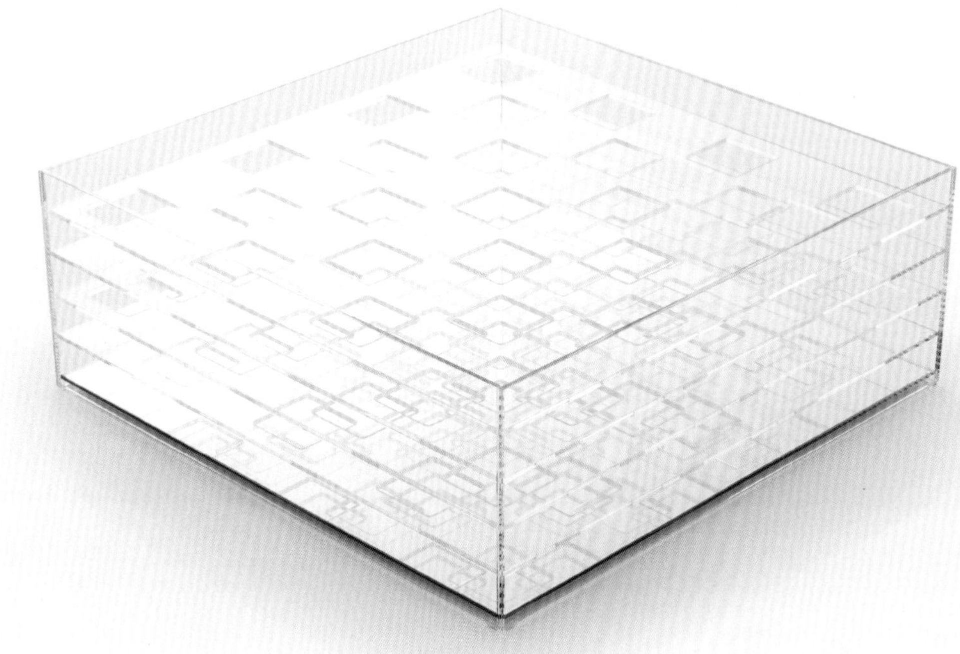

Window

Glas Italia

Normally, holes represent something absent. However, in this table the holes take on a presence – and look as though they are floating inside its glass surfaces. Using water jet-cutting technology, square holes were incised in a grid formation in each of the table's four glass panes, which were then inserted in a glass box. The holes' positions shift successively, layer by layer, and the bottom of the interior is mirrored, creating a pattern which changes according to the user's perspective.

492

Border Table

Eye of Gyre, Tokyo

A collection created for a solo exhibition held during Tokyo Design Week 2015. When designing furniture for mass production, it is difficult – if not impossible – to visualize all the circumstances in which those pieces will be used. The resulting designs inevitably tend to rely on an 'average specification' that can respond to various scenarios, and when filled with such furniture the space will often grow monotonous in use.

Therefore, the goal with this exhibition was to trial a new relationship between space and furniture by conceiving the design of the latter from the standpoint of a very specific space. Following an assessment of the gallery space, an unusual design process was embarked upon in which inspiration was found in elements that are normally 'troublesome' – such as the corner of a room, or columns protruding into the space. Two factors needed to be considered in the development of each design: first, how it could be structurally supported by the space; and second, how the space influenced the proportions of the furniture itself. This resulted in the creation of a series of tables constructed from 5-millimetre (1/4-inch) -square metal rods supporting small tabletops with a radius of 100 millimetres (4 inches). This design process utilized the element of space as a direct part of its deliberations by latching items, in an almost parasitic way, onto the corners of walls, and the edges of floors and exhibition stands.

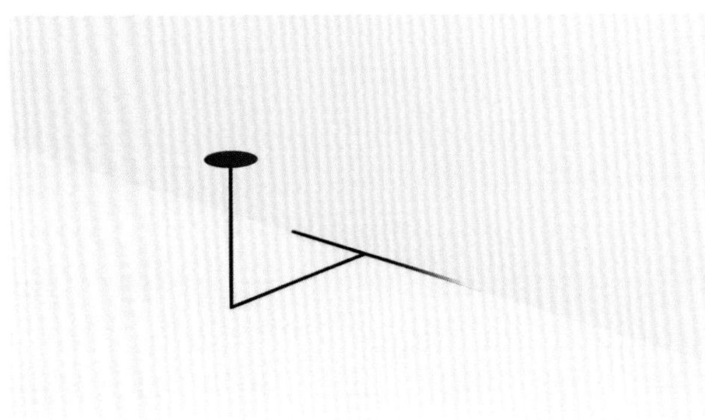

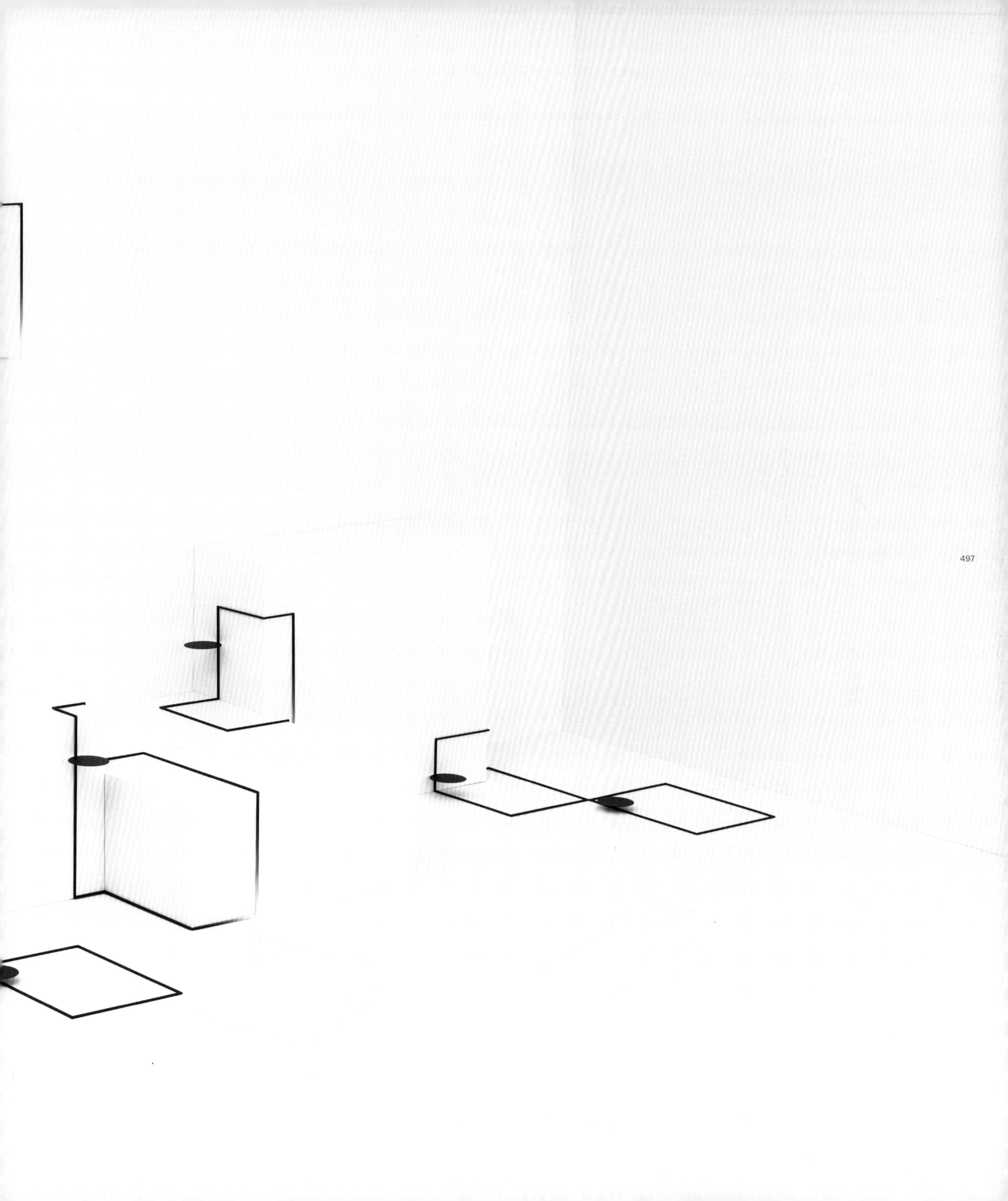

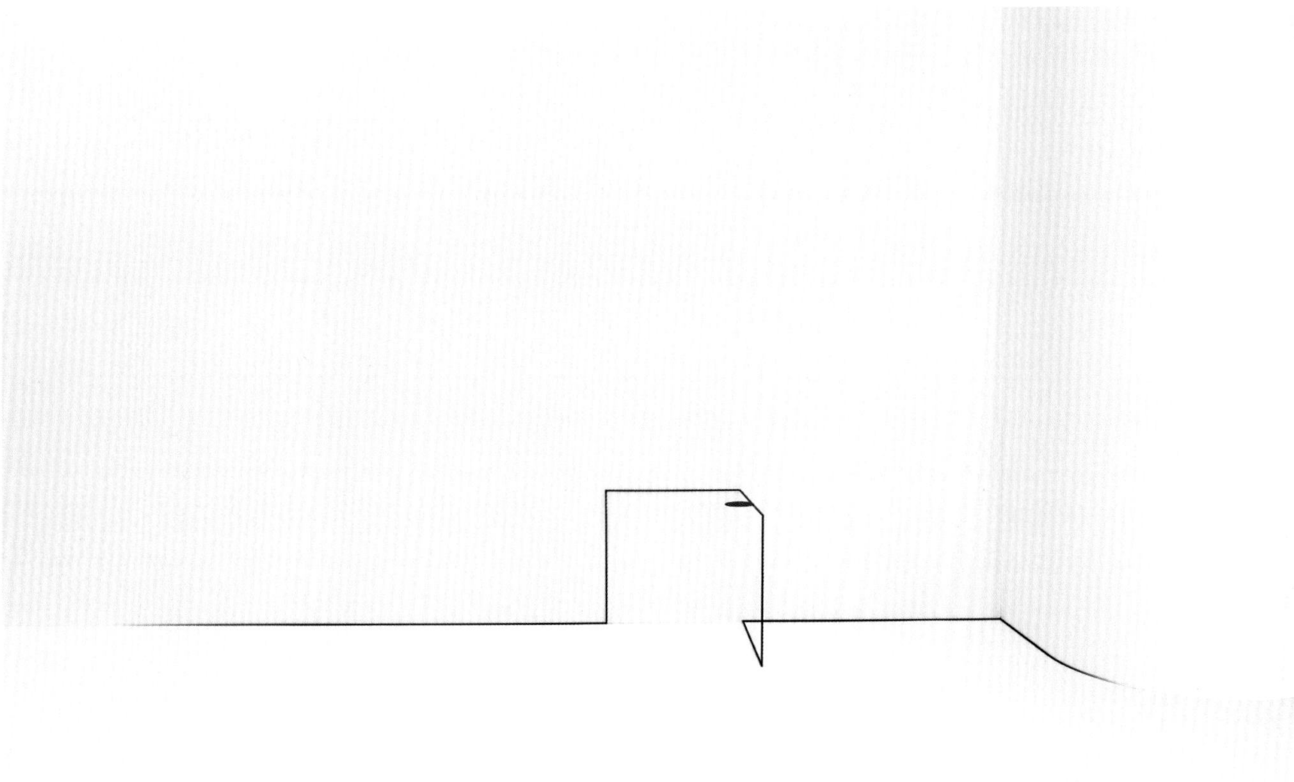

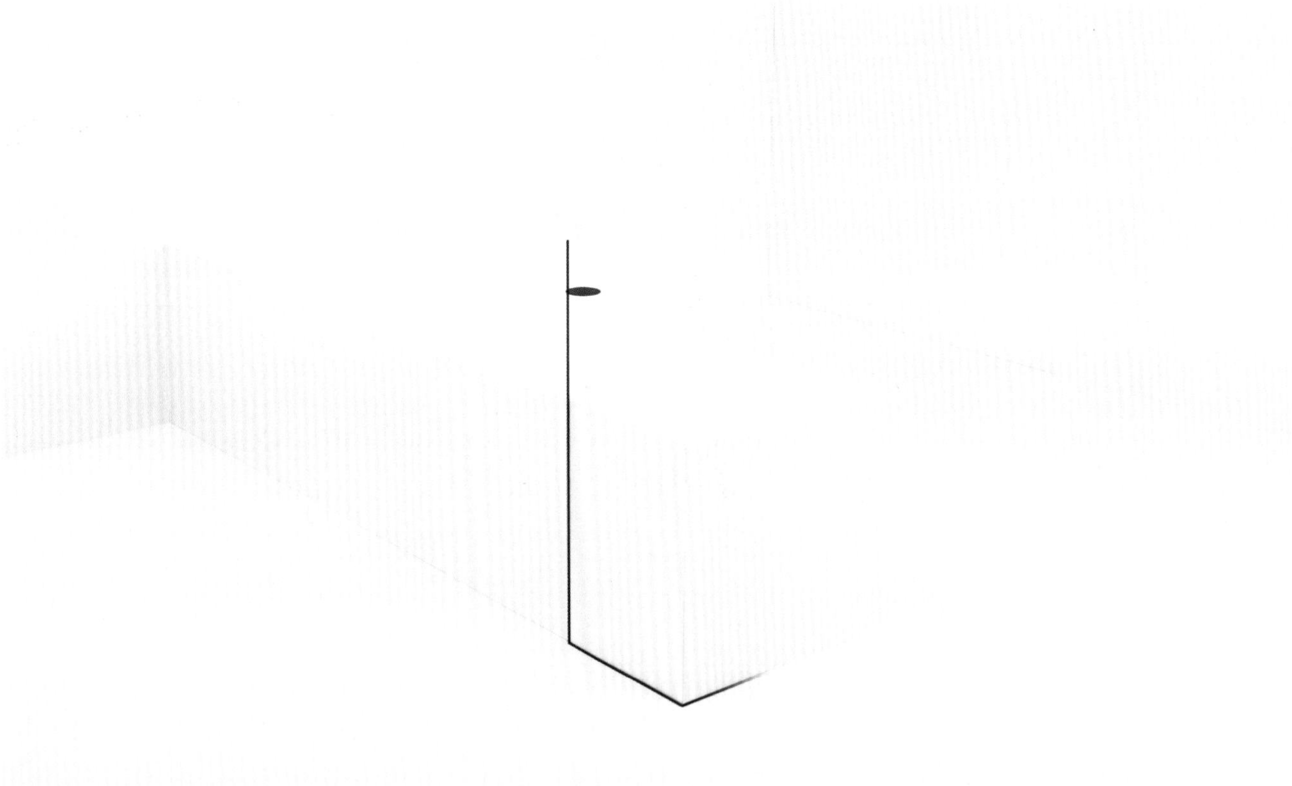

500

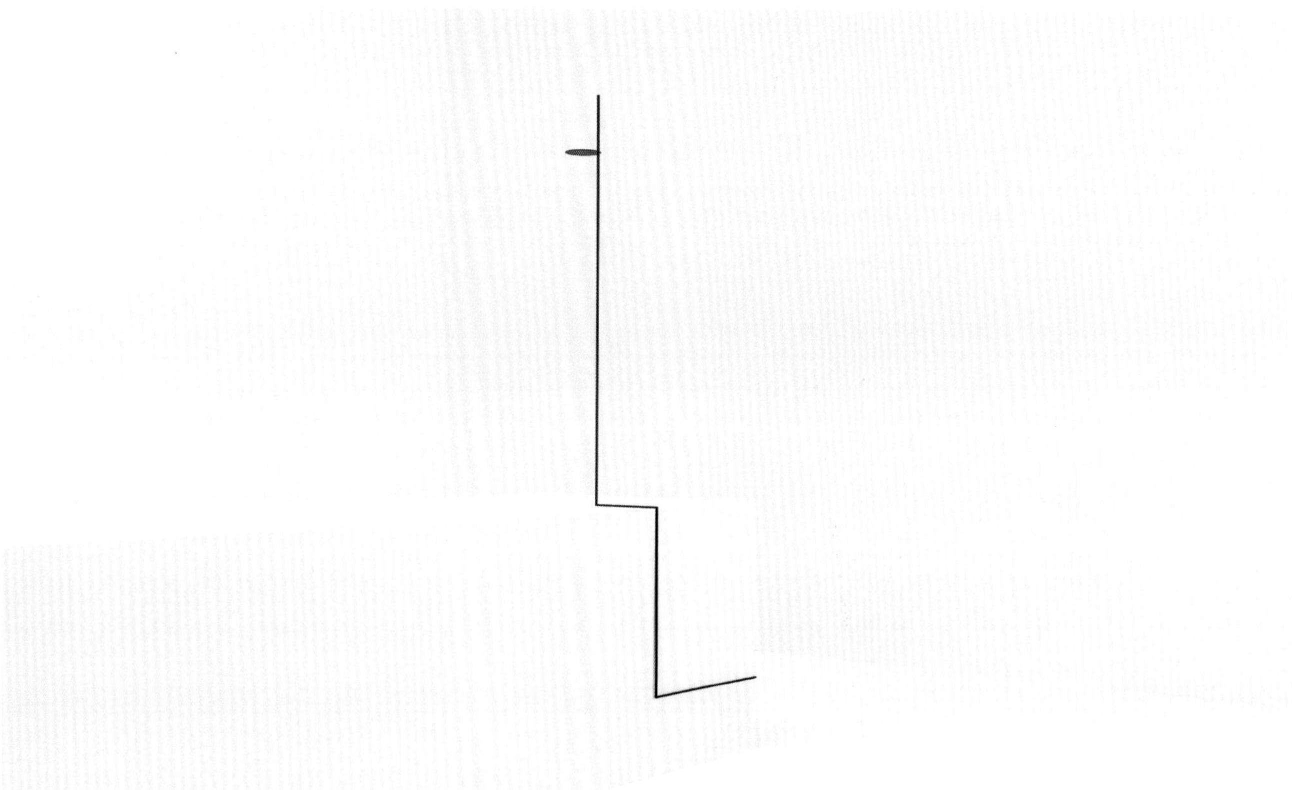

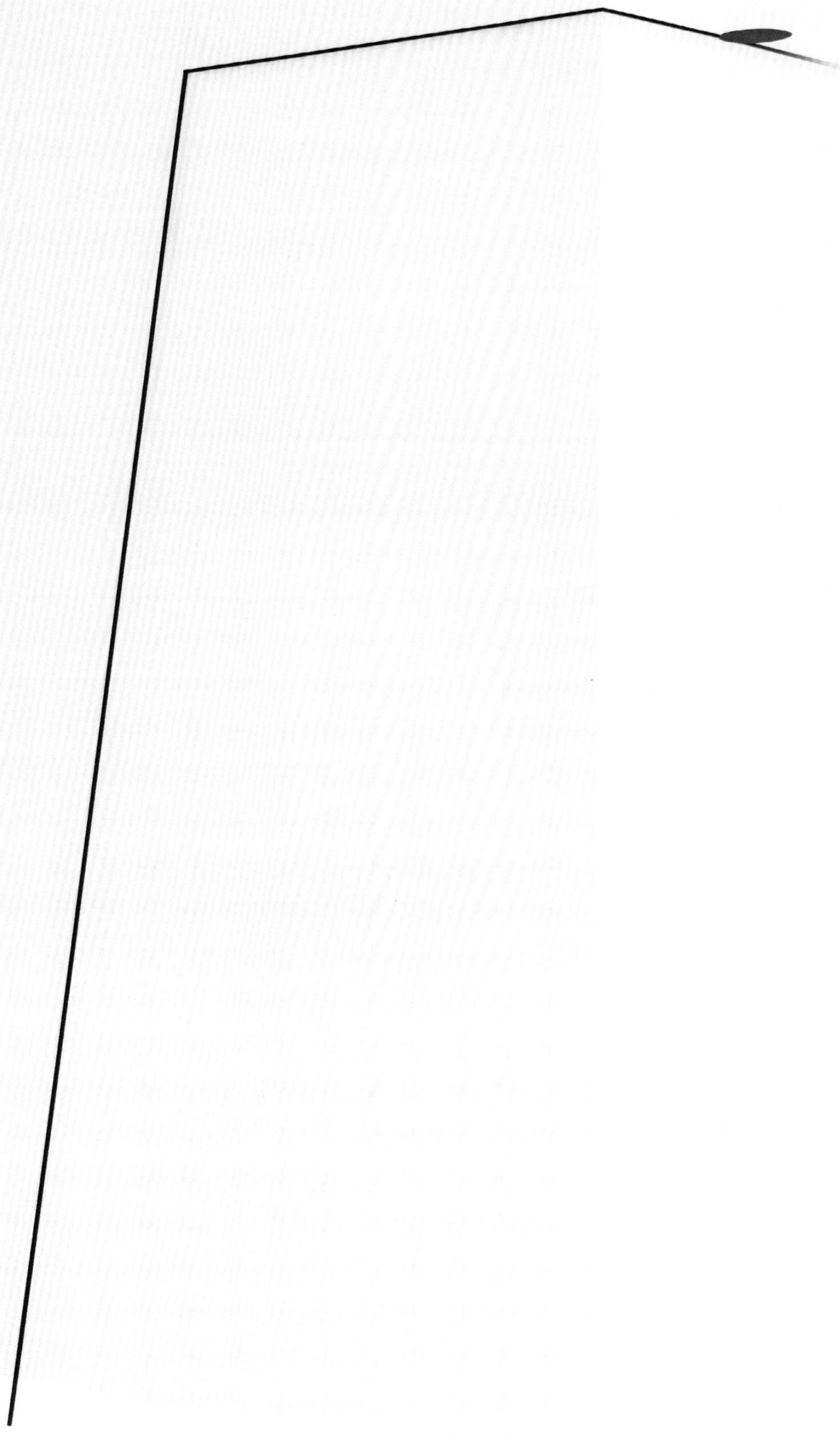

Catalogue of Works

2013

Pebble
Italesse

A collection of bowls shaped like river pebbles whose edges have been smoothed and rounded by the rushing current. They decrease progressively in size, just as pebbles become smaller as the water wears them away. When three bowls of different sizes are stacked together, their rim heights align so that they appear to have been carved from the same stone.

Ski
Swedese

In order to make a coat rack from the most basic elements possible, three legs taken from a single piece of timber were attached at their bases to create a structure like a very tall stool. When their upper ends were 'split' to form coat hooks, the legs resembled ski poles, upended and leant together for balance.

Curve
Berga Form

A stacking chair and bench that gently bend in the middle to create the curve of the seat. Constructed from a set of evenly spaced steel pipes, the collection was carefully designed to visually soften the otherwise hard, cold look of the metal furniture.

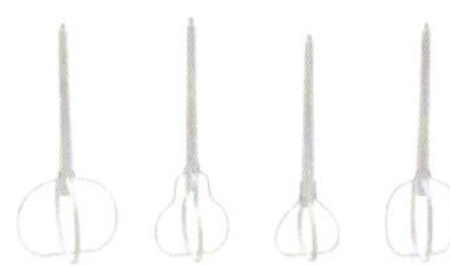

Clip
Discipline

A candle holder that uses steel frames to 'clip' the candle in place. Designed in four different fruit-like forms, the set is accompanied by a wooden tray that serves as the 'fruit bowl'.

Komori
Bisazza

A chandelier reminiscent of a colony of bats hanging upside-down from a tree branch or the ceiling of a cave. Evenly spaced glass shades, blown by Murano glass artisans, enclose LEDs and cast their light downwards. More support rings can be added to create a large 'colony', or removed to make a smaller one.

Stone-Edge Table
Caesarstone

A set of side tables designed to showcase the surface strength of Caesarstone, a material made of grained quartz. Rather than ostentatiously using it for the entire product, it was employed to protect the corners and feet of wooden tables – those parts most likely to be damaged in use. The seven different tables vary slightly in size and proportions, and each employs a different type of the material.

N Bottle
Hidetoshi Nakata

A packaging design for N, a high-quality sake produced by former Japanese football star Hidetoshi Nakata for the long-established brewery Takagi Shuzo. Created in a limited edition of 900 bottles, N's exclusivity enhances the worldwide allure of Japanese culture. The design brief requested a bottle that shielded its contents from sunlight and had an unconventional shape. The response was a minimally branded, matte-black cylinder like a stick of charcoal.

The cap was made by spinning tubular aluminium on a lathe, and two slight dimples were set into either side of the bottle for ease of pouring.

Join
Serralunga

A single planter that looks like an assemblage of five planters of different sizes. This effect minimizes its visual presence, and shows off the large-scale rotation-moulding technology that Italian manufacturer Serralunga specializes in.

Lump-Bowl/Lump-Cup/
Lump-Plate
Japan Crafts Oshima

Japan Crafts Oshima is a traditional Yamanaka lacquer workshop founded in 1909 in Ishikawa Prefecture. The discipline includes various techniques, but its workshops are known for their wood turning. The Lump collection takes advantage of this process to heighten users' awareness of the myriad qualities of the wood. The bases were enlarged and rounded so that users cradle their 'lump' – a tactile experience. The bowls' exteriors have a natural finish that showcases the wood grain, in contrast to the matte lacquer finish inside.

Oke Cup/Oke Carafe
Nakagawa Mokkougei

A collection of drinking items in Japanese cypress that utilize the barrel-making skills and techniques of Nakagawa Mokkougei, a traditional cooper whose wooden products have long been beloved of exclusive Kyoto inns and restaurants. Here, the wooden slats are bound with one hoop rather than the usual two, the walls are thicker than usual and nearly imperceptible spaces exist between slats in a new construction that disperses stress around the structure. Using one hoop achieves a clean look, and the thicker wood has a strong, comforting feel against the lips when lifted to the mouth.

Lid-Container/Lid-Vase
Japan Crafts Oshima

A series of bud vases and small containers created using traditional Yamanaka lacquer techniques. The lids, taken from the form of the traditional Japanese lacquerware soup bowl, were dramatically enlarged in order to entirely cover the smaller vessels beneath. This change in scale emphasized the small moment of delight that comes from taking off a lid.

Semi-Wrinkle Washi
Taniguchi Aoya Washi

Taniguchi Aoya Washi is Japan's only producer of three-dimensional washi, a process which is understood to be made by passing 3D moulds through plant pulp and then removing it once the pulp is dried. Adding a vegetable – *konnyaku* – creates wrinkles that enhance the paper's character but also interfere with its traditional seamless finish.

This design bridges the divide, creating lighting fixtures that combine both processes. The wrinkles are applied gradually, and they shrink the size of the fixtures such that final dimensions are difficult to predict. This hybrid process creates paper that combines softness with the tensile strength unique to three-dimensional *washi*.

Hanataba
Matsukan

Round chopsticks are slippery to use, but overly square-cornered ones aren't as comfortable to hold. Exploring ways of increasing the surface area of chopsticks in the hand in order to improve holding comfort, the design for Hanataba uncovered the tactile advantages of the pleated cross section. When viewed from the end, the chopsticks look like flowers, so a bunch of them kept together in a cup turns into a *hanataba* (bouquet).

Ume-Play
Gen-emon Gama

Gen-emon is one of Japan's preeminent porcelain kilns, dating from 1753. The Ume-Play collection reinterprets one of its best-known motifs: a plum flower known as *ume komon*. The design was cut up and rearranged to distil Gen-emon's hallmark blue and white underglaze, while creating a range of new pattern variations, which can be used together without breaking visual unity. The project took advantage of a new technology that enables the creation of very fine lines – a charcoal-based resist method known as *sumi hajiki*.

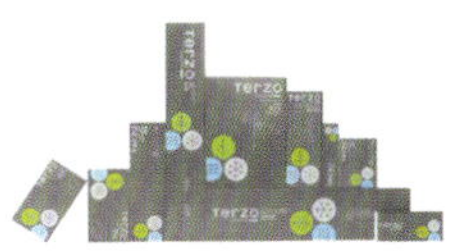

Terzo Branding
PIAA

An art direction project for Terzo, a brand of car roof boxes. Typically, car-accessory styling adopts an aggressive, outdoor image, but recognizing that younger buyers are less interested in the vehicles themselves, the design used clean lines and neutral tones to tap into their more modest lifestyles. *Terzo* ('third' in Italian) signified that this was PIAA's third brand. It was further divided into three thematic areas – 'winter' for ski/snowboard equipment, 'marine' for surfing and 'mountain' for cycling/camping – with a logo that combined the thematic colour and icon for each one.

Symmetrick
PIAA

A series of car roof boxes that are symmetrical along both axes. Their dual symmetry increases functionality: the carrier can be loaded with either end facing forwards, allowing its lengthwise opening to be positioned on either the left or right side.

2014

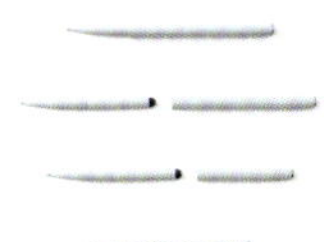

Flip Pen
Mark's

A pen that mostly consists of its cap. Depending on which end it is placed, the cap determines how the pen should be used – as a traditional writing tool, or as a tablet stylus. It also comes with a shorter cap, for easy portability.

Cross Pen-Stand
Mark's

Slender, cylindrical desktop penholders seem to be the default design. However, their shape allows only one way of use. The Cross Pen-Stand suggests an alternative design, which enables each pen to be stored separately, as well as accommodating flat objects, like rulers, cards or memo paper. The slots that enable this function form a repeated cross pattern when viewed from above.

Dot Envelope
Mark's

A traditional Japanese gift-giving envelope featuring the customary bicolour, decorative tie – but here, it is made entirely of dots printed on the envelope itself. The paper is embossed and treated with UV ink for a glossy, tactile finish, resulting in a simple, contemporary quality that fully expresses the giver's congratulatory feelings.

Circle Tag
Mark's

Sticky notes that take the shape of segments in a pie chart. Traditional square and rectangular sticky notes can be easily ripped off when their corners are caught on other sheets of paper. Increasing the sticky surface area and reducing the number of vulnerable corners on the sticky notes to the minimum resulted in this practical wedge form.

Peel Pen-Case
Mark's

A pencil case that can be unzipped, and its relatively rigid leather remains stiff enough to form a prop for the pouch – a case and penholder in one. A practical design for the office or on the go.

Kura Beer
Sekinoichi

Packaging designs for a craft beer produced in support of the Sekinoichi sake brewery in Ichinoseki – a city devastated by the Great East Japan Earthquake of 2011. The beer has three varieties: golden (*kin*) pilsner, red (*aka*) amber ale and black (*kuro*) stout.

Inspired by the unusual re-purposing of a *kura* (sake brewery) for the manufacture of beer, the products were named Kin Kura, Aka Kura and Kuro Kura. The *kanji* (logographic characters used in Japanese writing) for gold, red and black were incorporated into the *kura* character along with its English transliteration, 'Kura', to create logos resembling family crests.

1 Week Salad
Toshiba

A first-stage concept for 'Clean Room Farm' – Toshiba's plan to turn a former semiconductor factory into a farm – that reuses a sealed, bacteria-excluding 'clean room' to grow, harvest and pack organic vegetables. Its long-wave fluorescent lighting, air-conditioning system that maintains temperature and humidity levels, infrared surveillance system that monitors growing conditions, and sanitizing system that sterilizes packaging perfectly suit the cultivation of salad vegetables. The refrigerated salad lasts the eponymous '1 Week' since it has no contact with insects or bacteria. To emphasize freshness, smoothie-style containers sport numbers in different colours and typefaces for each of a possible thirty-one (monthly) days. These numbers indicate the best-buy dates.

Fusion
BoConcept

Fusion is a thirteen-piece furniture and homewares collection, designed for Danish firm BoConcept. The aim was to create new functions and visual effects by fusing together two or more different elements.

The sofa comprises four rearrangeable, differently sized cushions on a daybed, while other products include a clock with scissor-hands; a coffee table shaped like a pair of spectacles; and a reversible candleholder for tea lights or candle sticks.

Basket-Container
Kanaami-Tsuji

A collaboration with Kanaami-Tsuji, a Japanese company specializing in traditional wire-netting techniques. The Basket-Container, carefully constructed from individually hand-bent wires, is supported by a thin metal frame, resulting in a slender piece of furniture that is useful for holding small objects – and ideal for placing in tight spaces. The legs are narrower than the eyes of the netting, allowing them to be slotted through the surfaces of other items in the collection and stacked for easy storage.

Envelope
Tod's

A men's shoe designed for iconic Italian footwear brand Tod's. Two materials combine in its upper part – soft leather for the instep vamp, and suede for the rear quarter and backstay – and they are connected with a detail resembling an old-fashioned envelope's string-and-washer tie. A rubber, deck-shoe-type sole emphasizes Tod's design lineage. Near-invisible stitching brings out the shoe's softness while inside, a highly breathable mesh lining adds lightness and a sense of comfort.

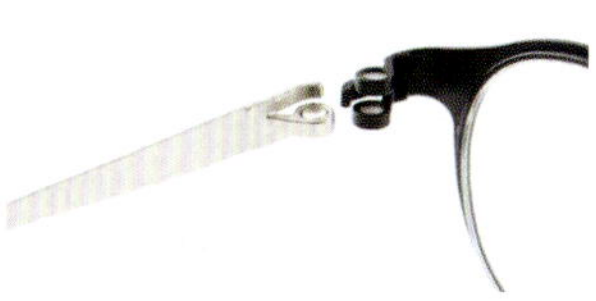

Magne-Hinge
Tokyo Optical

A pair of eyeglass frames with hinges that employ magnets rather than the usual screws, making its temples removable and the frames more flexible and difficult to break. The temples release smoothly when pulled from the side, so that users can create a variety of looks by purchasing them in different colours and materials.

Snap Glasses
Filston Global Design Works

A pair of reading glasses with a light, pliable polycarbonate-resin frame whose folded temples can be snapped onto the nosepieces. This feature creates a width of just 8.7 mm (less than 1/2 inch) at its narrowest point. The glasses' cloth case is a hollow cylinder, so that the spectacles can be inserted from either end – and its central hole ensures that the temples can still be secured in place.

Giftote-Bag
By | n

A design that extends the momentary delight of receiving a gift into the everyday by giving wrapping paper a 'second life'. After unpacking a present, the recipient can pass the ribbon through four grommets to transform it into a simple tote bag. It's easy to adjust the length of the handle, depending on where you tie it, and gift givers can choose from six different colours of ribbon. In order to allow more people to use it, costs were kept as low as possible by minimizing production steps and sewn parts.

Ribbon
Grasse Tokyo

A bottled room fragrance decorated with a ribbon. The ribbon not only identifies the fragrance inside but also functions as its diffuser. Taking advantage of the material's natural ability for absorption, the design turns an otherwise decorative element into something functional.

Sudare
Sugita Ace

A slatted furniture collection for compact exterior spaces. The high-backed chair, lounger and table fold flat when not in use, allowing them to double as partitions – like traditional Japanese *sudare* bamboo blinds – when placed before a railing or other vertical surface. The table's height and dimensions are easily varied by moving the position of its legs and adjusting the number of slats used on its top surface. Matching accessories, including trays and containers, have 'arms' that slot between the slats, for attaching to the furniture wherever needed. The frames come in five colours and the slats in two, for mix-and-matching.

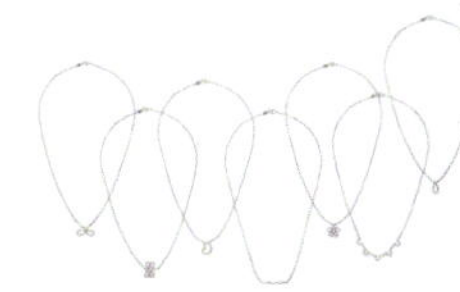

Single-Stroke
Vendome Yamada

A necklace collection made up of 10-carat gold chains. Each item is hand-brazed (i.e. fixed by soldering with a high-temperature alloy of copper and zinc) in order to harden the chain so that it forms the outline of one of a series of seven shapes – a dewdrop, heart, curling ribbon, flower, double loop, horizontal line or five stones. Overall, the collection consists of seven different necklaces, all of which appear to have been drawn by hand in a single stroke.

Dandelion
Takata Lemnos

A wall clock that, instead of using numerals, tells the time through the number of seeds depicted on a dandelion seed head. The stylized, line-drawn seed heads increase in quantity from one to twelve as they mark the hours around the clock face. The design alludes to time flowing onward like blown seeds dancing lightly upon the wind.

Tsuya Skin
Shu Uemura

A packaging design for a skincare range developed to achieve smooth, highly moisturized skin in its users. The product bottles express this ambition through their soft and slightly rounded edges, and the curving contours that line their exteriors – a subtle reference to the rippling liquid inside. Drawing on the notion that beauty comes from within, the collection is refined and elegant in its appeal.

Cover-Brella
Moon Bat

A lightweight, folding umbrella weighing only 108 grams (4 ounces) thanks to its carbon-fibre frame. Its cover is tucked into the normally 'dead space' of the handle, so – instead of being easily misplaced, like conventional covers – it is deployed on demand to rapidly encase the wet umbrella. Cover-Brella comes in eight different colours, all with a surface coating that reduces UV-ray penetration by 90 per cent compared with standard models.

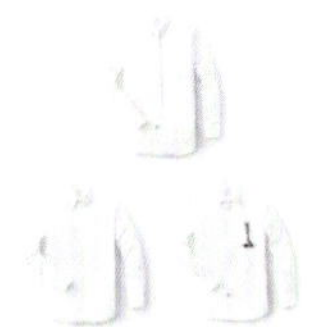

Fuse
Criminal

A men's shirt collection born from the concept of gently fusing items frequently found around shirts with the garment itself. Such things include accessories like neckties and sweaters worn with shirts, with one design featuring a pre-formed tie knot and necktie integrated into the shirt's collar. In other instances, details were added to accommodate products often found close to shirts, such as glasses and pens – for example, a 'bottomless' breast pocket for hanging one's glasses, with a wiping cloth sewn into it. These subtle customizations might seem ordinary at first glance but reveal their sophistication when viewed more closely.

Village
Häagen-Dazs

A design for an ice-cream cake to be sold at the Häagen-Dazs shop on the Champs Élysées, Paris, over Christmas. Since the festival prioritizes spending time at home with the family, the idea was to create a small cake-village featuring closely clustered milk-chocolate houses as decoration, which were filled with fudge and caramelized nuts. Their rooftops have varying patterns and are dusted with white-chocolate powder, so that the village appears nestled quietly in the snow.

Customers purchasing more than one cake receive a chocolate 'bridge'. These allow several cakes to connect with each other – turning them into larger tabletop 'villages', or a series of hamlets. To facilitate this, the surface of each cake undulates, landscape-like, in at least two directions, allowing the cakes to fit together easily regardless of how they are combined.

2015

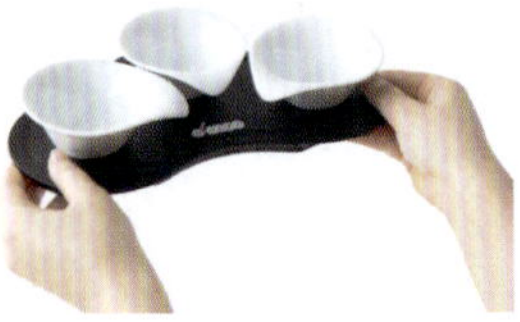

Tumamigui
Sushiro

A complete redesign for the conveyor-belt sushi restaurant Sushiro. Aiming to create a more compact, urban format, the project called for new name, logo, graphics, cutlery, uniforms and interior. To streamline and customize ordering, a new, tablet-based system was introduced to give customers more options with regards to portion sizes and add-on items such as sauces and toppings. The self-service cutlery and plates are flower- and petal-shaped, inspired by the idea of feeling free to 'pick and eat' – *tumamigui* in Japanese – whatever one wishes. These motifs are also used throughout the store interiors.

Totte-Plate
Miyama

Cooking utensils such as pots and frying pans have handles, but tableware such as plates and bowls do not. These items do have handles, *totte* in Japanese, giving them new functions: they are easier to carry, can be hung and stored on hooks, and are easier to hold when hot. This not only results in new ways of use but also provides them with a strong sense of durability as a tool.

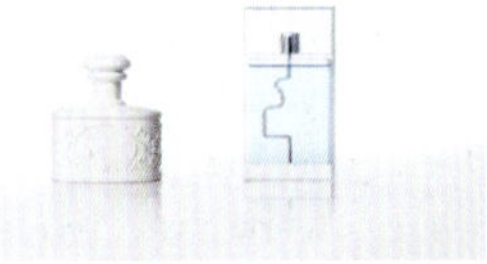

Fandango
Be Open

A perfume bottle designed for the group exhibition *The Garden of Wonders*, held in Orto Botanico Brera, Milan's historic botanical garden. Participating designers were tasked with redesigning a defunct perfume brand – in this case, 'Fandango' by Koehler, which gained popularity in nineteenth-century Russia. To express the ornate form of the original, the new design featured a tube inside a transparent bottle, whose shape mimicked the outline of the first edition. The tube came in two variations: one, a gradation of cold colours; the other of warm. They were respectively named Fandango -12.3 °C and Fandango +23.1 °C, referencing Moscow's average winter and summer temperatures.

IHI Company Ad
IHI

IHI is a Tokyo-based heavy-industries manufacturer with particular expertise in energy, social infrastructure, industrial machinery and aerospace. Dealing predominantly business-to-business, IHI is rarely noticed by ordinary consumers – and the fact that it supports Japanese daily lives in a variety of different ways is inevitably little known.

This advertising strategy for the company was based solely on its logo: the letters 'IHI' in blue font – originally suggested by their resemblance to the steel I- and H-section beams used in construction. Just like structural steel, duplicated logos were used ingeniously to 'construct' buildings, bridges, highways and vehicles.

Slide-Case
Hiroan

Hiroan, in Tokyo's Asakusa district, is one of Japan's oldest, small leather-goods workshops. Its unique *betahari* method layers two pieces of leather, tanned to 0.5 millimetres (less than 1/32 inch), creating extremely thin, tough and supple products.

Hasegawa Hiroshi, Hiroan's owner, and creator of its 'ultimate' thin-leather goods, says it is best not to overfill them. In recognition of this minimalist philosophy, Slide-Case's compact wallet and card case feature a subtle slit opening that encourages users to carry only the essentials – while the A5-sized organizer cover closes by sliding a pen into its clasp. To access stored cards, users need only to push a tab at the base of the wallet.

Beauty Library
Nature's Way

A composite store comprising a specialty boutique for organic cosmetics and a cafe. Shoppers can leaf through information on various products using a tablet, or on their mobile phones via QR codes, and casually try them out as if sampling a menu – while enjoying an actual menu of freshly prepared food and refreshments. Items are displayed on rows of shelves throughout the space, like stacks of books in a library, while individual 'study' desks provide space for testing the products on sale.

Marker-Shoes
Pioneer

A pair of leather shoes whose patterns are drawn directly onto their calfskin with markers. Unlike with dye or paint, the soft, fine calfskin absorbs the artificial marker ink in moderation, for a finish as transparent and limpid as watercolour. The shoes come in three designs each for the men's and women's collections, and in three colours for each design.

Stream
Asahi Woodtec

A flooring design for Osaka manufacturer Asahi Woodtec. By creating less conspicuous seams between the lateral joints, the grooves along the lengths of the floorboards are emphasized, creating a stronger sense of direction. These grooves come in three different interval patterns: wider ones expressing 'slowness', narrower ones 'quickness' and the third one a midpoint between the two. They can gently orientate occupants towards a particular view, or make one feel as if he/she is moving at a faster pace, such as in a corridor.

Architect Bag
Tod's

A bag that changes format according to its contents. In its extended form, it can hold A3-size drawings; folding it in half creates two spaces in which A4-size drawings and documents can fit; and the open, middle part of its flexible section houses rolled-up drawings or bulkier items. With the ability to store away the two handles in outer pockets, it can also be turned into a clutch bag. Drawing on the theme of 'a bag for architects', the project explored how focusing on specific functions can inform the shape of the end product.

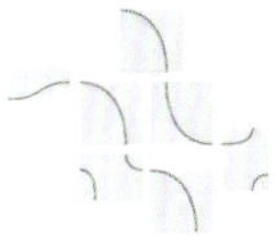

Path
Glas Italia

A table that combines supporting 'legs' of vertical, continuous curved-glass sheets with a flat glass top. The edges of the curves are black, resembling a path. When numerous tables are positioned side by side, this 'path' – seen from above – can continue seamlessly from one to the other.

Clear Shadow
Glas Italia

A dining table with a top made of two layers of glass – the upper, transparent layer is printed with a near-invisible pattern and the lower layer is white, to emphasize any shadows. The white glass sheet actually has two coatings: a matte-white top coat prevents the reflection of light, and a black undercoat stops the transmission of light. Although the print on the table's transparent layer is almost imperceptible, its presence is indicated in its shadows, which emerge on the white surface below – as do those of objects placed on the tabletop.

Ultime8
Shu Uemura

A packaging design for the 'beauty oil' product that forms part of the Ultime8 range from cosmetics brand Shu Uemura. The collection is named after the formula's characteristic combination of eight botanical oils. Ultime8's simple, cylindrical bottle, in an orange sepia tone with a dark-coloured cap, incorporates an intriguing drape-like profile that mirrors the flowing movement of the rich oil, and is intended to evoke the feeling of cashmere-soft, luminescent skin. The design also features a near-seamless integration of cap and bottle, as well as a gradation of sepia and darker colours to represent the blend of various oils.

Kinchak-Pouch
Aswave

A pouch that features a small *kinchaku* (drawstring pocket) on its side – a device that has been used in clothing accessories for centuries. Perfect for stowing small items that might otherwise end up lost at the bottom of a bag, the pocket is also ideally suited to use as a pencil case.

Abe Kogyo Showroom
Abe Kogyo

A showroom design for Abe Kogyo, a manufacturer of wooden doors, partitions, fixtures and custom-made furniture. Doors were set in 'walls' arranged in a radial pattern, as if the walls themselves were doors, opening on an imagined central hinge. The resulting arc, which also represents a door's trajectory as drawn on blueprints, was adopted as a motif – appearing throughout the showroom. This strategy not only integrated the temporary walls into the design but also introduced movement. Door types, and product names/numbers were displayed on the outer edge of each 'wall'.

Split-Tote
Aswave

A normal tote bag, with its paired handles, tends to lack stability when slung over the shoulder – and can also slide off. Conversely, a tote bag with just one handle has stability when carried over-shoulder but can be awkward to take things in and out of. Accordingly, Split-Tote divides a single handle springing from one end of the bag into two, secured on its front and rear faces – addressing both aforementioned shopping dilemmas and adding convenience.

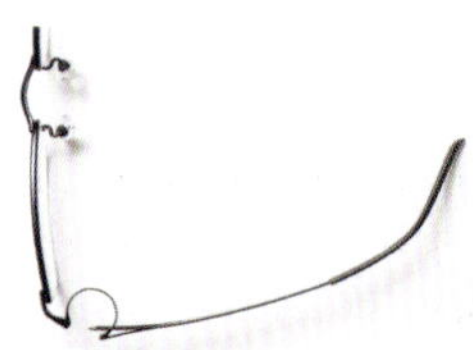

Zerogra-Fold
Meganetop

The Zerogra glasses are a leading product of Megane-Ichiba, an eyewear retailer that has more than nine hundred stores across Japan. Moulded from curved thin titanium plates, and with no hinges on either side, the eyeglasses are lightweight and offer superb wearability. To improve the limitations of the previous model – which, due to its single-body construction, could not be folded inwards – the new design adopted a reverse-hinge structure that enabled the frame to close fully. Additionally, to relieve pressure on a single point, two slits were incorporated into the end of each temple, around the reverse-hinge, which evenly distribute the load over the entire frame. The result was a small and compact design, with increased comfort and durability.

Hat Cloud
Sogo & Seibu

A renovation for a women's hat retailer located on the second floor of the Seibu Shibuya department store. The space is positioned in a slightly secluded area of the shopping centre, and has several downstand ceiling beams, testifying to a history of different uses. To unify this environment, cylindrical structures with 'mini-ceilings' at various heights were irregularly positioned like huge, pale-pink 'clouds'. In visual contrast, but continuing the cloud theme, individual hats were displayed like spindly umbrellas. The mid-section of the space became a seating area, with hats descending from the ceiling on longer hook-like variants of these umbrella handles.

Key to Style
Sogo & Seibu

An interior design project for a women's fashion floor within the Seibu Shibuya department store. The space connects by walkway to the 'Compolux' floor that nendo designed around the theme of a European park in 2013. The aim was to create a dynamic, casual setting targeting the younger customer, while maintaining the Compolux connection. The resulting concept was for a 'mobile amusement park' often held within such bigger parks, with sales areas inspired by circus tents and brand collections arranged in 'stalls'.

U-Note, Sa-Note
Mark's

Two ruled notebooks whose lines progressively fade towards the bottom of the page until finally disappearing. The unruled, plain white area is perfect for drawing sketches or illustrations. U-Note features fewer lines and is geared towards those whose work is driven more by intuition and feeling, while the greater number of lines in Sa-Note make it suitable for more logical thinkers. The naming comes from the Japanese pronunciation of 'right' and 'left', with the amount of space occupied by lines corresponding to a more right-brained (*u-no*) or left-brained (*sa-no*) style.

Pet-Pearl
Vendome Yamada

Our pets bring us joy and make us happy. In this jewellery collection, a small, single pearl plays the part of a pet; it 'resides' in a hamster wheel or a kennel made out of silver, or an acrylic aquarium. It is hoped that by wearing this jewellery the owner will receive the same kind of happiness as that given by a real pet.

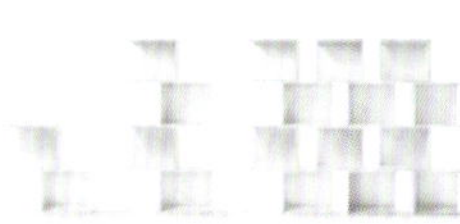

Chain
Hem

A modular shelving system designed for Hem, a German company offering online furniture sales. The units form half-cubes with two open sides, and are connected vertically by plugging in metal clips from the rear, allowing them be piled up with slight offsets. The components can be locked together snugly or space may be left between them – there is also the option of extending vertically, and combined units can even be shaped to fit along a curve or into the corner of a room.

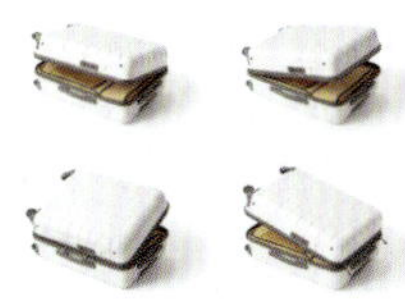

360
Ace

A polycarbonate suitcase with a 360-degree zip fastener that allows it to be opened from any direction. An interior, soft-shell lining increases the strength of the outer case, keeps rain from entering through the zipper and enables ample storage space without a centre-split construction. The design also provides easy access to separators and pockets even when in an upright position. With features including silent casters, external ribbing for increased rigidity and ultra-durable zip fasteners, the 360 model is the first release from Proteca since their rebranding in 2015.

Proteca
Ace

An interior design for the Proteca suitcase brand's first flagship store. The space was fitted out with stacking fixtures inspired by the wooden boxes and palettes used in transport, which travel the world on aeroplanes and boats – just like suitcases. While the in-store shelving and display stands were made of timber, a glossy-black acrylic coating was applied to the top of the palettes and the inside 'walls' of the box-shelving were painted white. These white, inner walls filled the boxes with soft light while also providing a neutral backdrop that drew customers' attention to the colours and textures of each product.

Connel Coffee
Sogetsu

A remodelling project for a cafe situated on the second floor of Tokyo's Sogetsu Kaikan building. The space boasts intact interiors from its original 1977 Kenzo Tange design so the redesign only involved the floor and furniture – its walls and ceilings were left untouched. In order to smoothly unify the partitioned space, nendo's 'Stream' boarding was fitted on the floor, the sides of the counters and the stairs – the joints on each being aligned to create the greatest possible sense of consistency within the room. Corian countertops were used to match the dark mirror finish of the ceiling and walls.

Index

Picture Credits

Unless otherwise noted, all images are courtesy and copyright © nendo.

Main Book

©'76,'19 SANRIO, Yuki Kawakami: 423; Daici Ano: 14, 23, 26–32, 42–5, 56–9, 62–5, 70–1, 76–7, 86, 96–7, 100–5, 108–9, 130–3, 146–9, 204–7, 210–17, 224–5, 236–41, 278 (r)–81, 298–303, 320, 344, 345 (t), 350–1, 365 (br), 400–1, 402 (l), 403–5, 426–7; Arketipo: 128; Joakim Blockstrom: 230–1, 315 (tr), 330–3, 418–19, 432–6, 474–5; Camper: 364–5, 365 (bl), 366–7; Cappellini: 66–7; Coca-Cola: 282; Jimmy Cohrssen: 87, 138–9, 156–7, 190–3, 338–9; Desalto: 384–9; © Disney, based on the Winnie-the-Pooh works by AA Milne and EH Shepard: 394–6; Emeco: 393 (l); Masayuki Hayashi: 15, 22, 38–41, 46, 52–5, 60, 72, 75, 80–5, 88–95, 98, 120–3, 125–7, 129, 134–6, 144–5, 158–63, 166–9, 174–87, 189, 198–203, 208 (l), 218–23, 226–7, 232–5, 272–3, 286–8; Hiroshi Iwasaki: 24–5, 140–3, 196–7, 208 (r), 209, 228–9, 244–8, 250–63, 266–71, 275–6, 278 (l), 283–5, 310–13, 337, 340–2, 352–6, 358–9, 362, 374, 376–81, 391, 406–7, 414–17, 420, 458–70, 473 (br), 492–501; Yoneo Kawabe: 137 (l), 150–5, 264–5, 290–1, 304–7, 336; Takumi Ota: 48–51, 390, 398–9, 402 (r); Kanako Sato: 10–13; Kenichi Sonehara: 444–7, 476–7, 480–1, 490–1; Koichi Torimura: 16–20; Ayao Yamazaki: 348, 368, 472; Akihiro Yoshida: 33–5, 61, 68–9, 73–4, 99, 137 (r), 172–3, 188, 292–7, 308–9, 334–5, 349, 360–1, 363, 369–71, 375, 392, 393 (r), 397, 410–13, 421–5, 430–1, 437–43, 448–57, 471, 473 (tl, tr, bl), 478–9, 482–9; Masaya Yoshimura: 274, 277, 314, 315 (tl, bl, br), 316–19, 321, 345 (b), 372–3, 408

Catalogue of Works

Daici Ano: 110 (c1, b), 111 (c3, b; c4), 112 (c3; c4, b), 113 (c2, c3, t; c4, b), 114 (c1; c2, t), 115 (c2, b; c4, b); 117 (c2), 323 (c1, m, b), 509 (c4, t); Arketipo: 322 (c3, t); Baccarat: 327 (c2, b); Berga Form: 502 (c2, t); Bisazza: 502 (c2, b); BoConcept: 504 (c4, t); Jimmy Cohrssen: 116 (c3, b), 322 (c2, b), 506 (c3, t); Council: 116 (c2, t); Discipline: 502 (c2, m); Driade: 324 (c3, t); Guzzini: 116 (c4); Häagen-Dazs: 506 (c2, b); Masayuki Hayashi: 111 (c3, t), 112 (c1, t; c2; c4, t), 113 (c1; c3, b), 114 (c2, b; c3; c4), 115 (c1, b; c2, t; c3; c4, t), 116 (c2, b; c3, t), 117 (c1), 322 (c1), 323 (c1, t; c3, t; c4), 324 (c1; c2, b); 327 (c1, b); Idea International: 323 (c2); Italesse: 502 (c1, t); Hiroshi Iwasaki: 324 (c3, b; c4), 325 (c1; c2, b; c3; c4, t), 326 (c1; c2; c3; c4, t), 327 (c1, t; c2, t; c3; c4); 502 (c3), 503 (c2, t; c4, t), 506 (c1, t); Kartell: 323 (c4, b); Lema: 325 (c2, t); Moroso: 322 (c3, b; c4, t); 325 (c4, b); Nacasa & Partners: 115 (c1, t); Takumi Ota: 116 (c1); 507 (c2, b); 508 (c4, t); 509 (c1, t; c4, b); Quodes: 322 (c2, t; c4, b); Rinascente: 326 (c4, b); Kanako Sato: 110 (c1, t; c2, b; c3, m, b; c4, t), 111 (c1; c2, t); Serralunga: 502 (c4, t); Kenichi Sonehara: 506 (c4, b), 507 (c4, b), 508 (c1, t, m); Swedese: 502 (c1, b); Tectona: 324 (c2, t); Koichi Torimura: 110 (c2, t; c4, b); 111 (c2, b); Ayao Yamazaki: 507 (c3, t), 508 (c2, t), 509 (c2, t); Akihiro Yoshida: 110 (c3, t); 502 (c4, b), 503 (c1; c2, b; c3; c4, b), 504 (c1; c2; c3; c4, b), 505 (c1; c2; c3; c4); 506 (c1, b; c2, t; c3, b; c4, t); 507 (c2, t; c3, b; c4, t); 508 (c1, b; c2, b; c3), 509 (c1, b; c2, b; c3, t); Masaya Yoshimura: 509 (c3, b)

Every reasonable effort has been made to acknowledge the ownership of copyright for photographs included in this volume. Any errors that may have occurred are inadvertent, and will be corrected in subsequent editions provided notification is sent in writing to the publisher.

Phaidon Press Limited
Regent's Wharf
All Saints Street
London N1 9PA

Phaidon Press Inc.
65 Bleecker Street
New York, NY 10012

phaidon.com

First Published in 2019

ISBN 978 0 7148 7813 3

A CIP catalogue record for this book is available from the British Library and the Library of Congress.

Commissioning Editor:
Virginia McLeod
Project Editor:
Robyn Taylor
Production Controller:
Lisa Fiske
Design:
Irobe Design Institute,
Nippon Design Center, Inc., Tokyo
(Yoshiaki Irobe and Tsugumi Arai)

Printed in China